The Future Highway Transportation System and Society

Suggested Research on Impacts and Interactions

Research and Technology Coordinating Committee (FHWA)

TRANSPORTATION RESEARCH BOARD
NATIONAL RESEARCH COUNCIL

National Academy Press
Washington, D.C. 1997

Transportation Research Board Miscellaneous Publication

Subscriber Categories
IA planning and administration
IB energy and environment

Transportation Research Board publications are available by ordering individual publications directly from the TRB Business Office, through the Internet at http://www.nas.edu/trb/index.html, or by annual subscription through organizational or individual affiliation with TRB. Affiliates and library subscribers are eligible for substantial discounts. For further information, contact the Transportation Research Board Business Office, National Research Council, 2101 Constitution Avenue, N.W., Washington, D.C. 20418 (telephone 202-334-3214; fax 202-334-2519; or e-mail kpeterse@nas.edu).

Printed in the United States of America

NOTICE: The project that is the subject of this report was approved by the Governing Board of the National Research Council, whose members are drawn from the councils of the National Academy of Sciences, the National Academy of Engineering, and the Institute of Medicine. The members of the committee responsible for the report were chosen for their special competencies and with regard for appropriate balance.

This report has been reviewed by a group other than the authors, according to procedures approved by a Report Review Committee consisting of members of the National Academy of Sciences, the National Academy of Engineering, and the Institute of Medicine.

This study was sponsored by the Federal Highway Administration, U.S. Department of Transportation.

Library of Congress Cataloging-in-Publication Data

The future highway transportation system and society: suggested research on impacts and interactions / Research and Technology Coordinating Committee (FHWA).

p. cm.

Includes bibliographical references.

ISBN 0-309-06218-7

1. Roads—United States. 2. Roads—Research—United States. 3. Transportation, Automotive—United States. 4. Transportation, Automotive—Research—United States. I. Research and Technology Coordinating Committee (U.S.). II. National Research Council (U.S.). Transportation Research Board.

HE355.F85 1997 97-48811
388.1'0973—dc21 CIP

Research and Technology Coordinating Committee (FHWA)

Raymond F. Decker, *Chairman*, USP Holdings Inc., Ann Arbor, Michigan
Don C. Kelly, *Vice-Chairman*, Jordon, Jones, and Goulding, Lexington, Kentucky
Allan L. Abbott, Nebraska Department of Roads, Lincoln
Richard P. Braun, St. Paul, Minnesota
John E. Breen, The University of Texas, Austin
William F. Bundy, Fleet Financial Group, Bristol, Rhode Island
A. Ray Chamberlain, American Trucking Association, Inc., Alexandria, Virginia
Forrest M. Council, University of North Carolina Highway Safety Research Center, Chapel Hill
Henry E. Dittmar, Surface Transportation Policy Project, Washington, D.C.
Nancy D. Fitzroy, Niskayuna, New York
Larry R. Goode, North Carolina Department of Transportation, Raleigh
Jean Jacobson, County of Racine, Racine, Wisconsin
Jack Kay, TransCore, Emeryville, California
Neville A. Parker, City University of New York, New York City
Gilbert S. Staffend, Farmington, Michigan
Dale F. Stein, Tucson, Arizona
C. Michael Walton, University of Texas, Austin
Richard P. Weaver, Big Arm, Montana
David K. Willis, AAA Foundation for Traffic Safety, Washington, D.C.

Liaison Representatives

Robert Betsold, Federal Highway Administration, U.S. Department of Transportation, McLean, Virginia
Francis Francois, American Association of State Highway & Transportation Officials, Washington, D.C.
Dennis Judycki, Federal Highway Administration, U.S. Department of Transportation, Washington, D.C.
Anthony R. Kane, Federal Highway Administration, U.S. Department of Transportation, Washington, D.C.

National Research Council Staff

Walter J. Diewald, Senior Program Officer, Transportation Research Board
Stephen R. Godwin, Director, Studies and Information Services, Transportation Research Board

Preface

The Research and Technology Coordinating Committee (RTCC) was established in 1991 to provide a continuing, independent assessment of research opportunities and priorities on which the Federal Highway Administration (FHWA) and other organizations can draw in developing their programs. In 1994 the committee published a report, *Highway Research: Current Programs and Future Directions* [Transportation Research Board (TRB) Special Report 244], that described the nation's highway research and technology (R&T) programs and identified several areas for emphasis into the next century. One area cited was the assessment of the role of highways within the U. S. transportation system. The report stated that "the nation's dependence on its highway transportation system is unlikely to change anytime soon," so "the kind of highway transportation system the nation wants in 20 years—or 30 years, or 40 years—should be identified and a research program that will help achieve such a system should be formulated." The committee believed this to be a topic of national significance warranting a substantial increase in research support by FHWA.

Reflecting on the many retrospective activities associated with the 40th anniversary of the Interstate Highway Act of 1956, several committee members noted that while the Interstate highway system has yielded many significant national benefits, it has also had some unexpected societal, economic, and institutional impacts. This report documents RTCC's efforts to provide FHWA with specific research recommendations designed to develop a better understanding of a broad range of societal, economic, and institutional factors that affect—and are affected by—the nation's highway transportation system.

ACKNOWLEDGMENTS

Much of the inspiration for this report came from Patricia Waller, Director of the Transportation Research Institute at the University of Michigan, who served on RTCC from 1991 to 1996 and who often urged the committee and FHWA to look beyond the "bridges and pavements" aspects of highway transportation and to focus some attention on the societal costs and benefits of the system that dominates this nation's passenger and freight transportation. Waller and other RTCC members who participated in the December 1994 FHWA Colloquium on the Social Costs of Transportation suggested that RTCC pursue an initiative aimed at providing FHWA with specific research recommendations related to the range of potential impacts of highway transportation (see FHWA for a copy of the draft document from the colloquium). Waller noted that transportation policy has costs and consequences related to both providing and not providing transportation and that these costs and consequences should be examined carefully as decisions are made about the system. This view became an underlying theme for RTCC's work on this topic.

Hank Dittmar served as the moderator for several RTCC discussions on the topic and led the task force that prepared four alternative future scenarios based on the committee's discussions. Members of the task force included Waller, Ray Chamberlain, Don Kelly, Gil Staffend, and Dale Stein [National Academy of Engineering (NAE)].

The study was performed under the overall supervision of Stephen R. Godwin, TRB Director of Studies and Information Services. The project director was Walter J. Diewald, who also drafted the final report under the guidance of the committee. Suzanne Schneider, TRB Assistant Executive Director, arranged the report review process. The final report was edited and prepared for publication under the supervision of Nancy A. Ackerman, Director, Reports and Editorial Services, TRB. Susan E. G. Brown edited the draft report. Martha W. Firestine edited the final report, and she and Marguerite Schneider prepared it for publication.

The committee would also like to recognize the work of those outside the committee and TRB who contributed to this effort. Ellen Cull assisted the committee as facilitator for several meeting discussions on scenario analysis. Sandra Rosenbloom of the University of Arizona, Thomas Horan of the Claremont Graduate School, and Charles Lave and Gordon Fielding of the University of California at Irvine participated in a roundtable discussion of the issues and provided helpful background information. Sandra

Rosenbloom, Mark Baldassare of the University of California at Irvine, David L. Greene of Oak Ridge National Laboratory, Lee W. Munnich, Jr., of the University of Minnesota, Daniel Sperling of the University of California at Davis, and Carol Zimmerman of Battelle Memorial Institute prepared background papers to inform the committee's deliberations. The papers are appended to the report for the interested reader. Whereas the interpretations and conclusions reached in the papers are those of the authors, the key findings of the committee appear in the main body of the report.

Ray Decker (NAE)
University Science Partners
Ann Arbor

Contents

1

Introduction

PURPOSE

Trends suggest that although more people will be driving more miles annually, by 2020 there will be little fundamental change in the current network of highways. With the Interstate highway system virtually complete and no new national road construction program under consideration, coping with highway congestion and its attendant delays will become a larger part of everyone's life by 2020. Yet rapid and continuing changes in the computer and telecommunications industries support the notion that there are new and exciting opportunities for changing the way in which people work, travel, and interact with one another. It is also possible that these and other factors—some that we can identify and some still unknown—will greatly affect the nation's highway system and its users by 2020.

The purpose of this study was to provide the Federal Highway Administration (FHWA) with specific research recommendations designed to develop a better understanding of a broad range of societal, economic, and institutional factors that affect—and are affected by—the nation's highway transportation system. The recommendations were prepared for FHWA by the Research and Technology Coordinating Committee (RTCC) on the basis of its view that certain issues are likely to be important to the highway system in the year 2020 regardless of changes in future travel behavior and the factors that determine it. RTCC believes that research on these issues, even though it represents a small portion of FHWA's overall research budget, has the potential for high payoff, particularly in view of the continuing influence of the highway system on individuals, communities, states, and the nation.

SCOPE

In its 1994 report, *Highway Research: Current Programs and Future Directions,* the committee cited the need for more information about the role of highways in the nation's transportation system and for research on the kind of highway transportation system the nation wants in the future (TRB 1994). The 1994 study addressed these needs, and as a result, it encompassed a broad range of topics, many of which are neglected in highway research today. It also yielded a wide range of recommendations, from specific research recommendations to suggestions that FHWA provide support or guidance to state highway agencies and metropolitan planning organizations (MPOs). This broad scope reflects the committee's view that FHWA's appropriate role in addressing the various topics covered in this report varies depending on how each topic fits FHWA's research mission. It also reflects the committee's view that FHWA needs to be an active participant in research on all these topics because of its overall role as the lead federal agency involved in highway transportation.

BACKGROUND

Past policies in support of the nation's highway system, particularly the Interstate highway system, have had a profound effect on the ways in which the nation conducts its business and its citizens carry out their daily lives. Some of the impacts of the Interstate highway system were unexpected, such as the ways that links between cities opened up areas for suburbanization and interchange locations became commercial hubs, drawing businesses from city centers. Moreover, such impacts were intensified by a booming postwar economy, public pressure to build more roads to meet rising travel demand, and a growing, more affluent population. The connection between human activity and highway transportation has become a close one as a result. For all its importance to the economy and modern lifestyles, however, this connection has not been explored extensively through research.

The persistent growth in passenger and freight travel demand is likely to continue because of such factors as the continuing suburbanization of residences and jobs, the increasing suburb-to-suburb travel, the growth of a flexible labor force with greater temporary employment, and the continued growth in household formation. FHWA's 1995 report to Congress on the condition and performance of the nation's highways describes such patterns of growth and suggests that highway travel demand will continue to increase at a rate of 3 to 4 percent per year (FHWA 1995). In addition, the aging population and increasing numbers of elderly drivers pose a new set of safety issues.

In its 1994 report on highway research, RTCC noted the broader implications of the highway system for society and called for more research that would take a long-term view of highway transportation and its interactions with other modes, land use, the environment, and the national economy (TRB 1994). The purpose of this research would be to better understand these interactions and to help shape the long-term direction of both urban and rural highway transportation in the United States as well as in regions and individual states. Such research could be viewed as supportive, or even as part, of strategic and contingency planning for the nation's transportation system and its major subcomponents.

Often drawing heavily on planning, management, and social science disciplines, this research might address issues such as the long-term relationship between transportation and economic development; travel demands in the 21st century; the opportunities, applications, and potential effects of new transportation and communications technologies; and sustainable transportation. Research funding in this category has increased over the past few years, but it remains a small percentage of total research and technology (R&T) spending. Although payoffs of this type of research are inherently difficult to measure, RTCC believes that such research is of crucial importance.

APPROACH

To identify a strategic set of research needs in this area, the committee set out to examine the societal, economic, political, environmental, energy, and other factors affecting and affected by the highway transportation system. The committee began its assessment by inviting several experts to participate in a preliminary roundtable discussion of key issues and trends that influence the transportation system. The committee then commissioned a set of background papers on the following key topics for review and discussion at a subsequent meeting:

- Travel, demographic, economic, and societal trends;
- Institutional, financial, and taxation issues related to how federal, state, and local governments are organized to provide transportation infrastructure;

- Communications and information technologies;
- Ecological, environmental, and energy-related issues;
- Alternative vehicle and fuel technologies; and
- Delivery of health and social services.[1]

The authors were asked to provide information about (1) the primary trends and driving forces in these areas, (2) the sources of significant changes in these trends or other technological breakthroughs affecting highway transportation, and (3) suggestions for needed research. In addition, the committee reviewed other material on trends and forecasts of important factors related to highway transportation, demographics, travel demand, vehicle technologies, and related topics.

A task force was then formed to carry out a scenario analysis based on the key factors that the committee identified as being most likely to affect the highway system. The committee chose to conduct a scenario analysis because it provided descriptions of alternative future scenarios framed by a set of key factors while avoiding undue reliance on trends. Predicting the future was not the aim, so the scenarios presented should not be mistaken for forecasts.

The committee then prepared its research recommendations on the basis of the issues, options, and questions suggested by the scenarios. As noted above, the committee recognizes that current research in these areas uses a very small portion of FHWA's R&T budget, funds for which there is considerable competition. The recommendations were prepared because the committee believes that even small efforts aimed at these topics can produce substantial dividends.

---

[1] One of RTCC's initial concerns was the transportation disadvantaged. The committee chose to focus its attention on the delivery of health and social services for several reasons. First, FHWA is currently involved in cooperative research with other agencies such the Federal Transit Administration and the Department of Health and Human Services—agencies that are responsible for assisting persons who may lack personal mobility and access. Second, although the Medicaid program currently incurs about $1.5 billion in transportation costs for Medicaid patients each year, ongoing changes in health care provision could result in even larger outlays as the number of patients grows and pressure mounts to provide more of these services in managed care programs. Finally, state highway agencies and MPOs are facing increased pressure to address the issues faced by those who lack personal mobility and access.

ORGANIZATION OF THE REPORT

Chapter 2 presents the committee's findings and recommendations. Chapter 3 presents summaries of the six commissioned papers that attempted to identify current trends and future projections of many factors that help define the nation's societal and economic activities and affect demand for the highway system. Chapter 4 provides a discussion of the four primary or driving forces selected by the committee and describes the scenario analysis undertaken by the committee. The report concludes with a discussion of several common themes in the scenarios that provided guidance for the committee's findings and recommendations. Finally, the papers commissioned by the committee to provide background information on current trends are included as appendixes.

REFERENCES

Abbreviations

FHWA	Federal Highway Administration
TRB	Transportation Research Board

FHWA. 1995. *1995 Status of the Nation's Surface Transportation System: Condition and Performance.* Report to Congress. U.S. Department of Transportation, Oct.

TRB. 1994. *Highway Research: Current Programs and Future Directions.* National Research Council, Washington, D.C.

2

Findings and Recommendations

This chapter presents the committee's findings and recommendations on the six topics identified in Chapter 1: future travel demand; roles and responsibilities of government; communications and information technology; ecological, environmental, and energy-related issues; vehicle and fuels technologies; and delivery of health and social services. These recommendations focus on research opportunities. However, the concept of research is interpreted broadly here to include not only specific research topics and projects that would be undertaken by the Federal Highway Administration (FHWA), but also coordination with other federal agencies conducting related research and more generalized support and guidance from FHWA to state highway agencies and metropolitan planning organizations (MPOs).

FUTURE TRAVEL DEMAND

There is a continuing need for more information about passenger and freight travel demand, especially considering the ongoing changes in employment and in demographic patterns and the overall effect of economic restructuring. In addition, the speed at which advances in communications and information technology are being adopted is affecting the ways in which businesses locate, manufacture, and ship products and the ways in which people make decisions about travel for work, shopping, and personal business. Such information is particularly important at the regional, state, and metropolitan levels, where transportation investment decisions are made. Although information about both users of the system and those who depend on public transportation is needed more quickly than it is currently available, there is considerable uncertainty about how to provide the information quickly and accurately in an affordable manner.

Most of the travel forecasting models currently in use were developed more than 25 years ago, but many changes in job locations, household formation, worker demographics, and the like have taken place since then. In 1992, recognizing the need to overhaul transportation modeling capability, FHWA together with the Federal Transit Administration, the Office of the Secretary of Transportation, and the Environmental Protection Agency initiated a research program to develop a set of integrated analytical and simulation models (TRANSIMS) and supporting data bases. However, microsimulation at the level of detail currently performed by TRANSIMS is possible only with very fast, high capacity computers not readily available to state and local highway agencies. While the committee recognizes the need for and supports such research, it concluded that state and local highway agencies need quicker, more accurate, and less expensive travel forecasting methods models to meet short-term needs.

Recommendation: The committee recommends that FHWA's planning research program give priority to developing analytical tools and methods for use by the states and metropolitan planning organizations (MPOs) for quickly, accurately, and inexpensively collecting and analyzing passenger and freight travel demand and trends.

ROLES AND RESPONSIBILITIES OF GOVERNMENT

The government roles in transportation at federal, state, and local levels are changing as a result of several factors, including the completion of the Interstate highway system, pressures to reduce the federal budget deficit, downsizing at all levels of government, and changes brought about by the Intermodal Surface Transportation Efficiency Act of 1991. Although the final outcome cannot be predicted, several changes in direction are identifiable: downsizing and devolution are likely to continue; states and MPOs will probably continue to exercise more responsibility for investment decisions requiring them to address a wider range of environmental, societal, and equity issues; and traditional public financing will be increasingly supplemented by public/private partnerships, privatization of some transportation services and facilities, and other types of innovative financing. Even though the most likely source of additional federal highway funds is an increase in the federal gasoline tax, recent congressional response to higher gasoline costs and current congressional proposals to turn the entire federal gas tax back to the states suggest that Congress is unlikely to raise the federal gas tax at this time to provide more highway funding. However, as highway users face increased congestion and delays on the nation's highways, pressure will build on Congress to increase the federal gas tax to pay for improvements and added capacity. The uncertainty surrounding changes under way in the institutional forms and responsibilities of transportation agencies at federal, state, and local levels raises questions about the future roles and responsibilities of these agencies at all three levels.[1]

Recommendation: FHWA should continue to assist its partners at the state and local levels by conducting research to estimate the economic benefit and development value of transportation. Such research can inform the public debate over increasing user fees to support transportation investments. FHWA should also conduct research aimed at understanding how the economic benefits of transportation investments accrue to the private sector and individuals. Such research can inform policy development regarding the appropriateness of public/private partnerships and private-sector financing of transportation facilities.

COMMUNICATIONS AND INFORMATION TECHNOLOGIES

As communications and information technologies change rapidly, there is a continuing need to monitor and understand them and their effects on pas-

[1] Institutional changes at federal, state, and local levels will probably also affect the roles of the professional and trade organizations and associations. These include the American Association of State Highway and Transportation Officials, the National Governors' Association, the National Association of Counties, the National Association of Regional Councils, and other associations that represent government agencies and the individuals in them. User groups such as the American Automobile Association and the American Trucking Associations and associations representing the construction and paving industries could also be affected.

senger and freight transportation supply and demand.[2] These technologies have already transformed freight transportation supply and demand through such innovations as just-in-time delivery, although some effects remain unknown and others are misunderstood. Moreover, factors as diverse as transportation deregulation on the one hand and changes in retailing and the structure of consumer goods delivery and distribution on the other have contributed to significant changes in freight demand characteristics as well as consumer behavior.[3] These changes—in combination with integrated logistics in freight management and the greater use of communications and information technologies—will continue to affect the performance of freight transportation, which relies heavily on the nation's highway transportation system.

Recommendation: With regard to freight transportation and communication and information technologies, research is needed to improve our understanding of existing and evolving impacts of both economic restructuring and information technology, as well as the potential impacts of future changes. Attention should also be given to whether communications and information technologies will reshape further the structure of selected consumer goods delivery and distribution and how this might affect both consumer travel demand for shopping trips and freight travel demand. Specific research should be undertaken to assess the impacts of the adoption of new vehicle, ITS, and highway design technologies on future freight travel demand.

Transportation decision makers need information on the potential impacts of new vehicle and intelligent transportation system (ITS) technologies on travel behavior and on the demand for new transportation facilities. These technologies also may affect the spatial distribution of businesses and residences, enable more people to work at home or at remote telework locations, and affect overall passenger and freight travel demand.

---

[2] A recent TRB study committee prepared TRB Special Report 246, *Paying Our Way: Estimating Marginal Social Costs of Freight Transportation,* which includes research suggestions for improving the understanding of freight costs as well as the policy implications of external costs and transportation subsidies.

[3] Significant changes in retailing include the increased number of "warehouse," "superstore," and membership retail stores, the increased volume of catalog shopping, and the emergence of "teleshopping."

Recommendation: Research should address the relationship between passenger travel and communications and information technologies, focusing on the information that travelers need and want to make travel and route decisions, the way in which technologies can best deliver this information to travelers, and the possible effects of these technologies on travel behavior. Research should also address ways in which the technologies might affect development within metropolitan areas: type and time of travel, amount of travel affected, and location of travel are all issues needing further study.

ECOLOGICAL, ENVIRONMENTAL, AND ENERGY-RELATED ISSUES

RTCC recently published a report entitled *Clean Air and Highway Transportation: Mandates, Challenges, and Research Opportunities,* which provides an array of priority research recommendations related to ecological and environmental issues (TRB 1997). The report includes recommendations for immediate research and for organizational reform; these are not repeated here. With regard to energy and highway transportation, RTCC believes that because the United States depends on foreign sources for more than half its petroleum consumption and because the nation's transportation system is almost solely dependent on petroleum, it is crucial that the nation be prepared for any interruption of foreign petroleum supplies. Contingency planning is essential if the nation is to successfully manage in time of crisis; although energy supply is not part of FHWA's mission, many state and local highway agencies will look to FHWA for information and guidance if such a crisis occurs.

Recommendation: As part of the work of its Office of Environment and Planning and its Office of Policy Development, FHWA should undertake research to examine how a crisis such as an interruption of petroleum supplies would affect the operation and maintenance of highways and other transportation facilities as well as the agencies responsible for them; it should continue to coordinate these efforts closely with the agencies more directly involved with petroleum supplies and emergency planning,

including the U.S. Department of Energy, the U.S. Department of Defense, and the Federal Emergency Management Agency.

VEHICLE AND FUELS TECHNOLOGIES

Although vehicle and fuels technologies are also outside its research mission, FHWA should continue its efforts to monitor changes in these technologies so that its research regarding mobility, safety, the environment, and highway financing is properly focused.[4] For example, changes in vehicle design and construction—size and weight of automobiles, for example, and increasing use of composite materials—need to be monitored and evaluated in terms of their effect on road design. Changes in the composition of the vehicle fleet—size, weight, and number of passenger vehicles and trucks; changes in the type and mix of vehicle materials; amounts and types of recyclable materials—should be monitored and evaluated in order to minimize the environmental impact as the vehicle fleet changes. Current research reassessing guardrail designs in view of the growing number of sport utility vehicles is one example.

Changes in vehicle fuels can have impacts on vehicle performance and highway financing as well as air quality; these changes should be monitored. Furthermore, proposals for new types of vehicles—such as super-lightweight cars that are far more energy-efficient and less polluting than current vehicles, small "commuter" cars that are more energy- and space-efficient, and "station cars" that would be available to transit users to complete their journeys—should be monitored and evaluated for their potential effect on highway design, highway funding, traffic operations, and traffic safety.

Recommendation: Although vehicle and fuels technologies are also outside its traditional research mission, FHWA should continue its efforts to monitor changes in vehicle and fuels technologies that can affect mobility, safety, the environment, and highway financing. FHWA, in partnership with the National Highway Traffic Safety Administration, needs to monitor and react to develop-

[4] See Appendix E for more information on current vehicle-related programs such as the Partnership for a New Generation of Vehicles.

ments that could affect the nature and safety of the passenger vehicle, as well as the integrity of the highway system.[5]

DELIVERY OF HEALTH SERVICES AND SOCIAL SERVICES

The committee noted that although access and mobility are very important to the delivery of health and social services, there is very little interaction between transportation policy making and health and welfare policy making.[6] As metropolitan areas expand and rural populations decrease, as demands on social services increase in urban and rural areas, and as health care delivery changes in form and location, social service decision makers could use more information about the costs and consequences of both providing and not providing basic transportation service. Moreover, the concentration of poverty in urban cores and isolated rural areas points to a broader social and transportation issue, inasmuch as most job growth is now occurring in the suburbs. At the same time, transportation decision makers could use more information about ways in which the transportation system affects the delivery of health and social services; and decision makers for health care and social services could use more information about whether adjustments to the transportation system can benefit the delivery of health and social services, particularly for those with limited access to either public or private transportation.

Recommendation: FHWA should take steps to assist in making the highway system more accessible to those who lack personal mobility. FHWA should cooperate with other federal agencies

[5] Although the aging population was a consideration in the committee's discussions, the committee chose not to include a specific recommendation about this group because of two reports that addressed specific issues of mobility and safety for older drivers and older pedestrians. These reports are TRB Special Report 218, *Transportation in an Aging Society: Improving Mobility and Safety for Older Persons,* and TRB Special Report 229, *Safety Research for a Changing Highway Environment.*

[6] There is increasing pressure on state highway agencies and MPOs to address the issues faced by those who lack personal mobility and access to service providers and to potential employers. Transportation officials lack information about these groups and their needs because in the past travel demand studies have concentrated on highway users, paying little attention to those who lack the ability to use highways. Given FHWA's mission, the committee believes FHWA should cooperate with federal agencies responsible for assisting groups who lack personal mobility and access to support the activities of state highway agencies and MPOs.

such as the Federal Transit Administration and the U.S. Department of Health and Human Services to support the development of analytical tools for states and MPOs to use in assessing the costs and consequences of providing and not providing basic transportation service in urban areas, suburban locations, and rural areas, particularly as new job opportunities increasingly are found in suburban locations far from center-city and remote rural concentrations of poverty. FHWA should also seek to understand the manner in which the highway transportation system affects the delivery of health and social services.

REFERENCE

Abbreviations

TRB Transportation Research Board

TRB. 1997. *Clean Air and Highway Transportation: Mandates, Challenges, and Research Opportunities.* National Research Council, Washington, D.C.

3

Review of Trends and Projections

Summaries of the Six Commissioned Papers

This chapter presents information obtained from the committee's efforts to identify current trends and future projections of factors that help define the nation's societal and economic activities as they relate to the highway system. Examining such trends and projections was the key to identifying a range of factors that affect the nation's highway system. The six commissioned papers including research issues identified by the authors are summarized in this chapter; the full papers can be found in the appendixes to the report.

FUTURE TRAVEL DEMAND: CURRENT AND EMERGING SOCIETAL PATTERNS[1]

The nation's population has grown 1.16 percent per year since 1980, with about 2.5 million people currently being added each year. The nation's population is aging; in 1990 more than 25 percent of the population was over 60, and more than 75 percent of those over 65 lived in metropolitan areas, most of those in suburbs. The elderly are the fastest-growing part of the

[1] The material in this section is summarized from Appendix A, which was prepared for the committee by Dr. Sandra Rosenbloom.

U. S. population; the number of people over 65 grew more than 20 percent between 1980 and 1990. In 1990 there were 6.2 million Americans over 85, a number expected to increase more than 400 percent by 2050. As younger drivers age, the increase in licensing among the elderly grows; within 20 years, most people over 70 will be licensed to drive.

Although the population grew 21 percent between 1969 and 1990, U. S. households grew almost 50 percent. One-person households increased 41 percent, single-parent households increased 36 percent, and married-couple-with-children households increased 8 percent. The population is becoming more diverse: Hispanics are expected to compose 23 percent of the population by 2050, while the white proportion drops to just over half.

Vehicle ownership continues to grow in the United States. Between 1969 and 1990 the average number of vehicles per household rose from 1.16 to 1.77, while households having 2 vehicles jumped 117 percent, an annual growth rate of almost 4 percent. The percentage of households without a car fell from 20 percent in 1969 to less than 10 percent in 1990. In 1990 only 6 percent of the entire population lived in a household with no car, compared with 21 percent in 1969.

The nation's work force and the work that it performs are changing. The participation of women in the work force increased more than 14 percent from 1970 to 1990. The number of married women in the work force rose from 33 percent in 1960 to 60 percent in 1990; working married women with children under 6 increased from 18 percent in 1960 to 60 percent in 1990. Between 1970 and 1990, the total number of service jobs grew 73 percent, while manufacturing jobs grew only 2 percent and agriculture jobs decreased 6 percent. Between 1982 and 1993, temporary employment increased almost 250 percent, compared with 20 percent for total employment. In 1993 there were 7.6 million telecommuters and an estimated 6 million workers who used their cars as their offices.

Population growth in the suburbs continues to outpace that in central cities. Suburban-population portions of metropolitan areas increased from 23 percent to 46 percent of total U. S. population from 1950 to 1988. In addition, since 1980 most employment growth has been in the suburbs; in 1990, 18 of the 40 largest job centers in the U. S. were located outside traditional downtowns. On the other hand, almost all neighborhoods characterized by extreme poverty are located in the 100 largest central cities, and population in the high-poverty census tracts more than doubled between 1970 and 1990.

Future Demographic Trends

By 2010, of the population over 65, more than half of all women and about 41 percent of all men will be over 75. The rate of new household formation will probably drop, with families headed by a woman alone increasing. Migration to the Far West has slowed, but there is still considerable migration to the South and West. While per-capita disposable income is predicted to increase at an average rate of 6.4 percent (or 1.5 percent when controlled for inflation) between 1992 and 2005, these gains are not projected to be distributed equally across the population.

In the coming decades the number of women employed outside the home is projected to continue to grow, with participation rates higher for black women than white women and lower for Hispanic women than white women. Retail trade is projected to replace manufacturing as the second largest source of total U. S. employment. Service-sector job growth is expected to be concentrated in highly skilled, high-pay jobs and in low-skilled, low-pay jobs (Silvestri 1993). The flexible labor force will probably continue to grow, but it is difficult to predict how many people will work at home in the future; the number of people able to spend at least part of their normal work week at home is expected to increase.

In addition to population growth and changes in the work force, travel demand will be affected by where people live and work. Suburban growth patterns are difficult to forecast, but growth is predicted in the new outer suburbs, near emerging edge cities, or immediately adjacent to inner-ring edge cities. The location of suburban employment growth depends on several factors, including whether suburban jobs concentrate at centers outside the traditional core or become dispersed in low-density suburban patterns. The author suggests that increases in land and construction costs in combination with land use regulations could lead to greater (relative) density.

Transportation Implications of Trends and Projections

As the population ages, the proportion of the elderly dependent on the automobile will probably increase, raising safety issues due to increased travel.[2] On the other hand, the very poor elderly often lack access to per-

---

[2] Travel projections are made more complex by uncertainty about the impacts of communications and information technologies on travel; these are discussed in a later section.

sonal automobiles and depend on public transit or some other means of transportation. As household size decreases, trips per household, trips per capita, and automobile trips per capita could increase. Female heads of household could experience longer work trips, a decrease in transit share, an increase in automobile trips, and an increase in reverse commuting. Women in the work force could experience increases in trips per capita, automobile trips, trip-chaining, and "serve-passenger" trips; a decrease in transit share; and more variability in trip scheduling and origin-destination patterns.

Absent other changes, continuing suburbanization of residences and jobs and a more flexible labor force point to potential increases in total and per-capita trip making, greater variability in travel schedules and origin-destination patterns, more off-peak trips, longer work and nonwork trips, more automobile and drive-alone trips, and a decline in transit use. Future travel will also probably involve more complex linked trips, more suburb-to-suburb trips, and more reverse commuting. Working at home could reduce the total number of work trips and the total distance traveled to work, but total trips could actually increase.

Research Issues Suggested by the Author

Several topics related to the aging population warrant research. They include research on whether changes to accommodate the aging population are needed in automobile and highway design and maintenance; traffic signs, signs, and information systems; and pedestrian facilities. Research is also needed to determine which changes will yield the largest improvement in highway safety. Research could also address how neighborhood design affects internal and external travel patterns.[3]

In view of the projected growing diversity of the population, research could address the basis of high transit use by immigrants and its potential for continuing. It could examine how variable employment patterns and the need for elderly caregiving affect travel patterns. Research could address the ways in which pricing and other incentives can affect traveler routing and scheduling decisions.

[3] For additional recommendations related to highway safety issues for older persons, see also TRB Special Report 218, *Transportation in an Aging Society: Improving Mobility and Safety for Older Persons.*

ROLES AND RESPONSIBILITIES OF GOVERNMENT: FINANCIAL AND TAXATION ISSUES[4]

The Intermodal Surface Transportation Efficiency Act of 1991 (ISTEA) led to many changes in the roles and responsibilities of government as they affect the nation's highway transportation system. Change continues as all levels of government face financial constraints. The private sector and consumers are being asked to play a greater role in transportation funding and policy. Pricing, technology, and privatization are all likely to have sizable impacts. Significant long-term changes in the financing of U.S. transportation infrastructure are shifting roles and responsibilities in financing and taxation for highway transportation.

Spending on transportation is shifting from the federal level to the state and local levels. Federal transportation spending represents about 1 percent of total federal expenditures; for states it accounts for 9 percent of the totals, and for local government, 8 percent of total expenditures. From 1982 to 1992 direct federal expenditures for transportation grew 3.1 percent annually, while federal transportation grants to state and local governments grew by 4.5 percent; the growth rate for direct state and local spending on transportation was 7.9 percent. From 1982 to 1992 state and local transportation spending per capita increased by about 20 percent, while overall growth in state and local spending was about 36 percent.

Transportation spending growth lags behind all other major spending categories of state and local governments. For example, over the past four decades, real per-capita transportation spending has remained unchanged while other areas of state and local spending have doubled, tripled, or quadrupled. Education represents about one-third of the increase, health and welfare about one-third, and the remainder can be attributed to public safety, environment and housing, interest costs, government administration, and the like. State and local government expenditures (adjusted for inflation) as a share of personal income increased from 12 percent in 1957 to 20 percent in 1992, while state and local spending on transportation dropped from 2.5 percent of personal income in 1957 to 1.6 percent in 1992.

Highway transportation and air transportation shares of total federal spending increased from 1980 to 1994, but other modes—transit, rail, and water transportation—lost shares. On the other hand, the federal share of

[4] The material in this section is summarized from Appendix B, which was prepared for the committee by Dr. Lee W. Munnich, Jr.

highway receipts dropped from 27 percent in 1960 to 19 percent in 1994, while the local share increased from 21 percent to 27 percent and the state share grew from 53 percent to 54 percent. Maintenance and noncapital expenditures grew from 38 percent in 1960 to 52 percent in 1993. Thus, while the burden of funding highways is shifting from the federal to the local level, more revenues are needed to meet highway system needs. In its 1995 condition and performance report to Congress, the Federal Highway Administration (FHWA) stated that to maintain 1993 highway conditions, annual average expenditures would have to exceed 1993 levels by 70 percent, and that to improve the highway system would require spending $65.1 billion per year, or 126 percent of 1993 outlays (FHWA 1995).

Automobile users are facing increased costs as well; the fixed costs of owning an automobile rose 31 percent from 1975 to 1994, while the real cost of oil and gas dropped more than 50 percent. Nevertheless, despite strong arguments to increase the federal gas tax, there is no political movement to do so. At the local level, where property and sales taxes make up the primary source of highway expenditures, there is considerable resistance to increases.

Downsizing and devolution are having impacts. Downsizing at all levels of government has achieved more streamlined organizations and reduced costs; its impact on service delivery is yet to be determined. Devolution is a long-term process of shifting powers and responsibilities from the federal government to state and local authorities; it is fueled by a growing distrust of government bureaucracy at all levels. ISTEA increased the flexibility of federal transportation dollars and moved surface transportation spending closer to block grant funding. Meanwhile, federal mandates that require additional expenditures at the state and local levels continue to confound the federal-state-local partnership. With greater decision-making responsibility acquired through ISTEA, states must deal more directly with environmental and social equity issues.

Potential Sources for Change

Innovation could result in changes at all levels of government. Technologies are available to more closely tie road use to user fees, and intelligent transportation systems (ITS) technologies have the potential to make better use of existing infrastructure if funding can be found to put them in place and operate and maintain them. Privatization, innovative financing, and partnering between transportation agencies and the private sector may provide the means to address specific problems or limitations in major corridors. Meanwhile, in response to efforts such as the National Highway User Survey, state departments of transportation are focusing their efforts on meeting user demands for higher quality in the construction and maintenance of highways (Opinion Research Corporation, 1996).

Competition and conflict between donor and donee states, between urban and rural areas, and between highway and transit advocates continue to cloud the picture of highway finance. Furthermore, with greater ISTEA-supported citizen involvement, future policy and funding issues will attract more participants to the decision-making process. Greater concern for environmental quality, sustainable growth, and public health and welfare could greatly affect the form of future transportation investments.

Research Issues Suggested by the Author

Research could address a range of issues regarding roles and responsibilities of the different levels of government. These include the new federal role in transportation in view of current fiscal constraints and the effects of a more privatized and market-based transportation system on transportation funding, economic efficiency, social equity, and roles and responsibilities of public transportation agencies. Finally, research could address which new partnership models are appropriate for transportation and how they can be promoted.

Other questions include whether full-cost pricing (and full-benefits analysis) can be applied in making transportation decisions at the state and local levels; how communications and information technologies can change the delivery of transportation services in terms of efficiency, service, and demand; the role of transportation policy in affecting commercial development patterns and densities; and determination of which infrastructure investments will contribute the most to productivity and global competitiveness.

COMMUNICATIONS AND INFORMATION TECHNOLOGIES[5]

Travel demand is affected by technology as well as economic, societal, demographic, land use and housing, and transportation policy factors. Cur-

[5] The material in this section is summarized from Appendix C, which was prepared for the committee by Dr. Carol A. Zimmerman with Christopher Cluett, Judith H. Heerwagon, and Cody J. Hostick.

rently, personal travel can be characterized by increased reliance on automobile travel, a flexible work force, more suburb-to-suburb travel, more dispersed job locations, and more trip chaining. Commercial freight movement can be characterized by (*a*) diminishing margins in a global economy, smaller lot production, and decentralized manufacturing base, which make logistics, timing, and transportation costs more and more important; (*b*) increasing nonstore retailing (one analyst estimates that by 2010 retail sales in stores will be 45 percent of the total as compared with 85 percent in 1992); and (*c*) changes in truck volumes and weights.

The relationship between communications and information technologies and transportation is evolving quickly. Communications and information equipment is increasingly more portable, powerful, and affordable, and consumer acceptance is high, although skewed toward consumers with high incomes and advanced education. New software technologies have improved acceptance and use, as have emerging improvements in interface technologies. New communications and information technologies can both substitute for and stimulate travel demand.[6] Trip elimination occurs when technology allows (*a*) people to replace travel with communication from geographically dispersed locations, (*b*) information to be sent electronically rather than physically, and (*c*) people to access documents and data electronically. It could mean telecommuting to work, watching pay-per-view entertainment and sporting events, using technology to gather product information before purchase and consumer information at the point of purchase, using remote classrooms, or accessing medical information and health care diagnostic services electronically.

Communications and information technologies can also stimulate travel by creating new jobs and higher disposable income, expanding the scope of interpersonal relationships, increasing expectations for rapid response, and providing mobile communications that encourage more travel by making it easier to work anywhere, including in the car. Finally, communications and information technologies can also stimulate the overall level of interactions, possibly leading to increases in both travel and communications.

---

[6] This summary focuses on the effects of communications and information technology on travel demand. The committee recognizes that most effects on travel and location are indirect effects of the use of such technologies. Examples include uses in transportation technology (e.g., global positioning systems and traffic information); for just-in-time production (e.g., automated inventory and shipping scheduling systems); and in job-related activities to support more use of flexible work schedules. A potential indirect effect of each of these applications of communications and information technology is a change in either freight or passenger travel demand.

Commercial freight movement has already been affected by communications and information technologies: automated weigh stations, toll collection, credential purchase concepts based on carrier-mounted devices, automatic vehicle location based on global positioning systems, and radio frequency–generated bills of lading have all been implemented successfully.

Finally, the growing need for repair and reconstruction of large sections of the Interstate highway system, particularly on urban routes with high traffic volumes, suggests that communication and information technologies might be useful to help reduce traffic disruptions and delays associated with the reconstruction activities.

Research Issues Suggested by the Authors

Research is needed on the impact of communications and information technologies on travel behavior, trip characteristics, and trip substitution and demand; on how providing travel information using different technologies affects travel decision making and mode choice; and on methods for measuring the effects of communications and information technologies on transportation supply and demand. It is also needed to determine the ways that communications and information technologies can be used to reduce traffic disruptions and delays associated with highway repair and reconstruction, particularly on urban routes with high traffic volumes.

ECOLOGICAL, ENVIRONMENTAL, AND ENERGY-RELATED ISSUES[7]

There are five keys to understanding transportation impacts on ecological, environmental, and energy issues:

1. The pervasiveness of the connection between human activity and transportation;
2. The difficulty of achieving full-cost pricing of transportation;
3. Government provision of much transportation infrastructure as a public good, amplifying the importance of public policy in addressing transportation's impacts on the environment;

[7] The material in this section is summarized from Appendix D, which was prepared for the committee by Dr. David L. Greene.

4. The potential for petroleum dependency (the U. S. transportation sector is 97 percent petroleum dependent) to become a significant economic problem in the United States at any time; and

5. The combination of (*a*) scientific uncertainty about the size and seriousness of environmental impacts and (*b*) the difficulty of establishing the value of largely extramarket goods, which makes it inherently difficult for society to decide how much of its resources to devote to solving transportation's energy and environmental problems.

Transportation is a large part of modern economies. The U. S. transportation system generated more than 10 percent of the nation's gross domestic product in 1995, producing 4.5 trillion passenger-mi and 3.6 million ton-mi of freight, while consuming 24.0 quadrillion BTUs. Highways account for 90 percent of the nation's passenger miles, about 25 percent of the intercity ton miles, and about 75 percent of transportation's energy use. Although the U. S. car and truck fleet continues to grow, the non-U. S. fleet is growing faster—3.3 percent a year as opposed to 7.4 percent a year. In 1950 the U. S. fleet (50 million units) was 70 percent of the world fleet; today the U. S. fleet (200 million units) is only one-third of the world fleet (BTS 1995b) The demand for transportation is also fueled by a population that is growing at a rate of about 2.5 million a year.

Growth in transportation energy use in developing countries has averaged 4.5 percent annually over the past two decades. Motorized transport and its energy use in Europe and Japan has been growing at a faster annual rate (1.8 percent) than in the United States (1.3 percent); globally expanding motorized transport produces about 20 percent of the world's greenhouse gases (GHGs).

A new automobile emits less than 10 percent as much pollution as a pre-1967 automobile; yet 182 of 268 metropolitan statistical areas failed to attain one or more of the National Ambient Air Quality Standards. Vehicle-related reasons for this failure include the 3 to 4 percent annual increase in demand for vehicle travel and associated increases in fuel consumption and the inability of emissions control technology to work as well in real-world conditions as in test conditions. Off-cycle operation, superemitting vehicles, and larger effects from air conditioner use than expected also increase air pollution. Options for reducing vehicle emissions include inspection/maintenance programs, greater use of reformulated gasoline, and increased use of alternative fuels.

Human activities contribute to GHGs, especially CO_2, which has increased from about 1000 million metric tons (MMT) in 1860 to 6200 MMT in 1991. Scientists also suggest that global temperatures have increased between 0.5°F and 1.1°F in the past 100 years. Predictions for future changes are extremely uncertain, but according to the author, solving the climate change problem could cost as much as $2,000 per capita per year in the United States.

World oil prices doubled in 1973 and again in 1979 because of actions taken by the Organization of Petroleum Exporting Countries (OPEC) nations, which currently supply nearly 50 percent of the demand and hold 60 to 75 percent of proven reserves. U. S. petroleum imports are currently within 1 percent of the historic high of 46 percent in 1978. U. S. dependence on OPEC for petroleum becomes an issue only if the price is raised or the supply reduced; without any changes in the status quo, consumers have little incentive to change their behavior or manufacturers to develop alternative sources of motive power.

Highways and streets use about 1 percent of the total area of the lower 48 states; the total of all built-up and urban land uses, including transportation, is about 5 percent. As land is used for transportation infrastructure, natural habitats are damaged in several ways: replacement by infrastructure, disturbance to adjacent habitats, fragmentation of habitats, and the direct killing of animals.

Several topics that reflect a common concern about energy and the environment are receiving increasing attention. The first is sustainable development, which is development that meets the needs for the present without compromising the ability of future generations to meet their own needs. The second is full societal cost pricing of transportation, which is aimed at requiring users to pay all direct and indirect costs of travel regardless of mode. The third is integrated transportation and land use planning, which aims at greater coordination in the implementation of policies that lead to land development at the regional level. These trends could combine to generate policies that would account for as many external costs and benefits as possible, particularly as communications and information technologies make such accounting possible.

Research Issues Suggested by the Author

Technology will be the major impetus for changing transportation's relationship to the environment by mitigating existing problems and developing alternatives. Potential solutions include new vehicle technology, new infrastructure for new energy sources, and new ways of managing highways, but considerable research and development (R&D) is needed before these are realized. Alternative fuel vehicles and vehicle efficiency improvements are possible, but conventional technology cannot meet higher CO_2 standards. ITS technologies can make transportation systems more safe and efficient; they can also provide the information needed to achieve fully internalized user costs. Full societal cost pricing and benefits determination, sustainable-development goal setting, and integrated land use and transportation planning all point to greater use of pricing mechanisms to manage transportation externalities. However, options such as raising the gasoline tax or adopting user charges for highway use are politically charged at this time, even if the revenues are used for construction or rehabilitation of the existing system.

VEHICLE AND FUELS TECHNOLOGIES[8]

Fuels have been modified over the years to reduce lead levels, increase octane, achieve climate adaptation, and respond to the needs of electronic fuel injection. The 1990 Clean Air Act Amendments accelerated efforts aimed at reducing emissions; nevertheless, more changes are likely in fuels and vehicles to achieve safer operation, lower emissions, improve fuel efficiency, and accommodate ITS technologies. Other potential changes include greater use of lightweight but costly composite materials.

Vehicle energy efficiency has improved about 30 percent from 1985 to 1995, but much of this gain has not been translated into fuel efficiency. Instead, consumer demand for larger vehicles with more power and fuel-consuming options has offset much of the fuel efficiency gain. With slowly expanding use of vehicles, total fuel consumption will also probably expand. This trend is unlikely to be altered much by the use of ITS, although safety and efficiency might improve. Furthermore, although ITS could bring improvements in paratransit and ridesharing, analysts note that the current trend away from transit use is unlikely to be affected substantially by ITS. Reducing fuel consumption can be achieved through technology changes, altered consumer preferences, and government action; however, the federal government has been reluctant to impose direct fuel restrictions or raise taxes to reduce fuel use. On the other hand, consumers currently show lit-

[8] The material in this section is summarized from Appendix E, which was prepared for the committee by Dr. Daniel Sperling.

tle interest in fuel economy because of low U. S. oil prices coupled with fading public concern for energy security and apathy toward climate change. Thus, most current trends point to increased dependence on private motor vehicle use.

Public demands for reducing oil imports, GHG emissions, and other environmental impacts such as noise have been muted; the desire to travel by automobile is strong and growing. Although the public appears to have lost interest in goals other than air quality, dependence on petroleum imports and global climate change could become more urgent as developing countries begin to use petroleum at a more rapid pace. Government intervention has been minimal; the exception has been the Clinton Administration's Partnership for a New Generation of Vehicles, launched to spur interest and investment by the U. S. automobile manufacturers in much more energy-efficient technologies.

Most metropolitan areas are expected to have achieved air quality standards within the next decade. Continued growth in population and vehicle use will forestall efforts by the federal government to reduce its commitment to air quality goals. California's continuing air quality problems and related initiatives will remain a focus for those supporting continued improvement in air quality.

In 1992 federal legislation set a goal of 10 percent market penetration of alternative transportation fuels by 2000 and 30 percent by 2010; requirements for government and private fleets were part of the legislation. Alternative fuels include ethanol from corn, natural gas, propane, methanol, and battery electric. The most effective alternative fuel for reducing GHGs is alcohols from cellulosic biomass such as corn and sugar cane. By 1995 less than 2 percent of vehicular fuel consumption was of alternative fuels. The alternative fuel goals (10 percent and 30 percent) are not likely to be achieved; moreover, congressional enthusiasm for fleet rules regarding alternative fuels has dissipated.

Battery-powered and hybridized internal combustion engine vehicles have many energy and environmental benefits, but the major automobile manufacturers have been slow to invest seriously in electric drive (E-D) technology (although General Motors Corporation recently offered its E-D vehicles for lease). California's zero emission vehicle mandate probably cannot be sustained outside California without other justifications. Concerns with cost and performance and lack of public commitment to energy and environmental goals could derail E-D initiatives; moreover, the more widespread use of batteries for power would introduce large amounts of materials into the environment, some of which may be toxic.[9] Although government support for E-D vehicles could spur the market for very small vehicles, smaller vehicles would also create safety problems.

Research Issues Suggested by the Author

The transportation community confronts four sets of issues related to different vehicle power sources and fuels: R&D, infrastructure design and investments, regulatory policy, and financing. Virtually all federal vehicle propulsion R&D is funded by the U.S. Department of Energy. The U.S. Department of Transportation (DOT) has practically no expertise in alternative fuels and E-D vehicles. DOT and FHWA recognize that road infrastructure and ITS deployment programs must be compatible with vehicle technology and fuels programs. However, FHWA should monitor how vehicle technology and alternative fuel programs might affect vehicle electronics capabilities, emergency services, and highway infrastructure programs. Regulations and design standards that could hamper changes in vehicle and fuels technologies need to be examined carefully; FHWA would have to coordinate its efforts with the National Highway Traffic Safety Administration, Environmental Protection Agency, and other affected agencies. Finally, as the market for alternative fuel and E-D vehicles grows, FHWA and the states should begin examining other methods for rational and equitable highway financing in addition to taxes on motor fuels.

DELIVERY OF HEALTH SERVICES AND SOCIAL SERVICES[10]

Several groups highly dependent on publicly provided health and social services are often constrained by their lack of personal mobility and access to the service providers. Particularly vulnerable populations include the inner-city poor, the elderly, the physically or mentally disadvantaged, immigrants, and single parents with young children. These groups often lack the money,

[9] Sperling (1995) states that improved technology and mass production will probably not bring the cost of battery-powered electric vehicles (E-Vs) down to that of gasoline cars. He also states that performance of E-Vs should not be expected to be equal to that of gasoline-powered cars.

[10] The material in this section is summarized from Appendix F, which was prepared for the committee by Dr. Mark Baldassare.

access to an automobile or public transportation, time, or personal assistance needed to travel long distances to obtain health and social services. Inasmuch as the diverse arrays of health and social services are usually housed in facilities dispersed over large and geographically widespread metropolitan areas, the problems posed by the transportation system can be daunting. These problems can be complicated further if service agencies consolidate to reduce costs or if the service area is expanded, resulting in longer trips for service.

Many factors are straining existing institutions and budgets; they include suburbanization, economic restructuring, immigration, aging of the population, an increasingly geographically dispersed and socially diverse elderly population, changing family roles and household composition, the persistence of an urban underclass, and increasing demands for health and human services in the suburbs. In addition, communities continue to adjust to reductions in the federal role in the provision of such services. The restructuring of welfare programs in the face of budget reductions can affect the accessibility of program services; the large number of political jurisdictions in metropolitan areas leads to further decentralization of services.

The nation's automobile-based transportation system exacerbates the problems of accessibility for vulnerable populations. Most metropolitan transit systems are designed to provide access primarily to job locations in the central business district and do not serve dispersed health and human service locations well. If health and human service locations become too scattered within a metropolitan area, service recipients are required to undertake more and longer trips to reach multiple locations; if locations are consolidated, then recipients may face longer trips. Increased travel to service locations could also increase the likelihood of accidents and injuries. The vulnerable populations using these services could be affected in the future by (*a*) reductions in funding for highways and public transit, (*b*) reductions in federal funding for health and human services, and (*c*) underestimation of the growth of such populations.

Suggestions for improving service and reducing costs include the following: locating health and human service facilities closer to highway and transit links; locating health and human service facilities closer to the vulnerable populations; improving public information about health and human services; and, possibly, instituting some form of transportation assistance for vulnerable populations. The current trend in health care is managed care with centralized facilities and an emphasis on outpatient care; this can add to the total number of trips. Those dependent on public transportation may find it difficult and time-consuming to seek care at such facilities. Welfare reform places strong emphasis on job training to help people get off welfare rolls; situating job training at locations not well-served by public transportation creates a hurdle for those without automobiles, a hurdle that can be heightened by child care needs and its transportation requirements.

Research Issues Suggested by the Author

Research could provide more information on the following topics: the impacts of suburbanization on the use of highways and the demands for health, social, and emergency services; travel demand related to health and social services; the travel demands being created by ongoing demographic changes; and the impacts of projected demographic changes and service trends on highway safety.

REFERENCES

Abbreviations

BTS	Bureau of Transportation Statistics
FHWA	Federal Highway Administration

BTS. 1995a. *1995 Transportation Statistics Annual Report.* U.S. Department of Transportation.

BTS. 1995b. *National Transportation Statistics 1996.* Report DOT-BTS-VNTSC-95-4. U.S. Department of Transportation.

FHWA. 1995. *1995 Status of the Nation's Surface Transportation System: Conditions and Performance.* Report to Congress. U.S. Department of Transportation, Oct.

Opinion Research Corporation. 1996. *National Highway User Survey.* Washington, D.C. May.

Silvestri, G.T. 1993. Occupational Employment: Wide Variations in Growth. *Monthly Letter Review,* Nov.

Sperling, Daniel. 1995. *Future Drive: Electric Vehicles and Sustainable Transportation.* Washington, D.C., Island Press.

4

Scenario Analysis

This chapter describes the way in which the Research and Technology Coordinating Committee (RTCC) used scenario analysis to develop its research recommendations. Scenario analysis has been used to assist decision making in the midst of high uncertainty (Wack 1985). Decision makers have found the technique useful as continued reliance on forecasts in a rapidly changing environment often misses critical turning points in the environment. The technique was suggested by several RTCC members recently involved in scenario analysis and the development of specific research recommendations based on scenarios (Bonnett and Olson 1994; Global Business Network 1994; FHWA 1996). The following section presents some basic information about the scenario analysis process and about how RTCC chose key driving forces and used them and scenario analysis to describe four alternative future scenarios. This is followed by brief descriptions of the four alternative future scenarios. A final section describes the common themes that emerged from RTCC's examination of these scenarios and that provided guidance for the research recommendations.

SCENARIO ANALYSIS PROCESS

Background

A scenario is a tool for ordering one's perceptions about alternative future environments in which today's decisions might be played out (Global Business Network 1994). The aim of scenario analysis is not to gather data for constructing a model but rather to systematically examine the available data to configure a number of different possible futures and then to examine the choices they present. Scenario analysis gives decision makers an opportunity to pretest concepts and decisions against a range of potential futures so that current decisions can be examined against such futures. Scenario analysis is not intended to predict the future or create a most desirable or most likely future.

Inputs to scenario analysis change over time and its outputs can be affected by the technical backgrounds and experience of the participants (Morrison, 1994). A basic step in scenario building is identifying driving forces that shape the future, in this case the driving forces shaping the highway transportation system's future. Driving forces are the factors and elements that influence a given state of affairs; they act as boundary conditions for the scenarios. Driving forces are often interrelated and complex, so considerable effort is made in scenario analysis to isolate those that are key to the subject of the investigation. The committee decided which driving forces were the critical uncertainties; four of these became the key driving forces.

It is impossible to account for all the uncertainties that define the future. Because scenarios are intended to organize a vast amount of information and translate it into a framework for judgment—in a way that no model can do—there is no correct set of key driving forces in scenario analysis. However, because the key driving forces are consistent across the set of alternative futures described, scenario analysis permits examination of the issues the futures emphasize. Although it is unlikely (but acceptable) that the future will actually reflect any of the scenarios, it is probable that the future will evolve within the range of possibilities that the scenarios suggest. The goal of RTCC was to identify common themes and issues of strategic significance within the alternative scenarios and to develop research recommendations in recognition of these themes and issues.

Selection of Driving Forces

Scenario analysis begins with a focal issue or decision; the issue RTCC addressed was the future state of the nation's highway transportation system. After reviewing the commissioned papers and other related documents, RTCC developed a long list of potential key factors or driving forces that could help describe the nation's future highway system (see box). With the aim of identifying a limited set of driving forces—those considered to be

1. Future travel demand:
    - Population growth, immigration, and diversity;
    - Industrial restructuring;
    - Flexible labor force;
    - Demographic discontinuity (baby boomers without replacement);
    - Suburban growth patterns;
2. Roles and responsibilities of government: finance and taxation trends:
    - People's willingness to fund transportation systems;
    - Privatization of transportation services;
    - Pricing strategies;
    - Competition for transportation dollars;
    - Downsizing and devolution;
3. Communications and information technologies:
    - Passenger travel effects: trip substitution, elimination, stimulation, temporal distribution;
    - Commercial freight movement: automation and productivity;
    - National Information Highway;
    - Computer and telecommunications equipment;
    - Public acceptance of and comfort with computers;
4. Ecological, environmental, and energy-related issues:
    - Public's valuation of energy, ecological, and environmental issues and extent to which this is connected to transportation;
    - Degree of certainty regarding external societal costs and benefits of transportation;
    - Dependence on petroleum-based fuel;
    - Public policy on transportation's energy and environmental problems;
    - Greenhouse gas emissions and global climate change;
    - Urban air quality;
5. Vehicle and fuel technologies:
    - Hybrid technology, including electric drive technology;
    - Energy and environmental concerns;
    - Consumer preference for vehicle amenities (performance, comfort, etc.);
    - Market drivers for efficiency breakthroughs;
    - Internal combustion engine improvements;

6. Delivery of health and social services:
   Level of funding and support for social services and care;
   Integration of social services and transportation industries;
   Market drivers for health services;
   Public perception of the connection between social services and transportation;
   Emergency services;
7. Other significant driving forces:
   Changing public values with regard to safety and security (in the contexts of vehicles, highway, and human beings);
   Public values with regard to right to access;
   Travel and tourism industry (related to leisure time, state promotion, private industry);
   Attitudes (public and political) toward long-term investment;
   Globalization of the economy.

the most important and most uncertain—the committee established criteria for screening its options. It decided that the key driving forces should (*a*) be critical to the future of the highway system, (*b*) be largely independent of the highway system, (*c*) have the potential for a major shift over the next 25 years, and (*d*) be characterized by considerable uncertainty. Applying these criteria to the driving forces, RTCC selected the following four critical uncertainties as its key driving forces:

1. Highway-related *technologies:* technologies related to highway planning, design, operations, construction, and maintenance; vehicle and road materials; vehicle design and operating characteristics; and in-vehicle communications and information systems;
2. *Finance:* funds available for financing transportation projects;
3. *Energy:* whether low-cost energy supplies would be available; and
4. *Environment:* whether environmental protection remains a major policy goal.

Although other potential driving forces, such as current demographic trends, are also critical to future outcomes, they are more likely to follow current trends.

Scenario Development

In scenario analysis each driving force has two possible outcomes or directions that in combination are used to define the range of alternative scenarios. The committee defined the two directions for each driving force as follows:

- **Highway-related technology 1 (HT1):** Highway-related technology is widely used; innovations continue to be supported, developed, and implemented.
- **Highway-related technology 2 (HT2):** Highway-related technology is not supported and stagnates; innovation is stifled.
- **Finance 1 (F1):** Public financing for highways is readily available as public demand for and political willingness to support mobility and environmental concerns come together in support of a higher fuel tax.
- **Finance 2 (F2):** Political willingness to use and increase user fees is lacking; public finance covers only maintenance (if that); most new facilities are funded by the private sector.
- **Energy 1 (E1):** Low-cost energy supplies are curtailed and the cost of energy increases dramatically.
- **Energy 2 (E2):** Low-cost energy is readily available.
- **Environment 1 (V1):** Public accepts environmental degradation as a genuine threat and supports attempts to protect and improve the environment.
- **Environment 2 (V2):** Public does not accept environmental degradation as a genuine threat and does not support improvements.

There are 16 possible combinations of four driving forces with two potential outcomes. After examining the 16 combinations, the committee chose 4 that it believed represented the widest range of potential futures with a minimum of overlap. These four combinations are presented in Table 4-1.

TABLE 4-1 Combination of Driving Forces in Each Scenario

Scenario	Driving Forces
1	HT2, F2, E1,V2
2	HT2, F2, E2,V2
3	HT1, F2, E2,V1
4	HT1, F1, E2,V1

DESCRIPTIONS OF SCENARIOS

Scenario 1: Scarcity and Downward Spiral

Scenario 1 is very bleak; each of the four driving forces is negative.

- **HT2: *Highway technology stagnates.***
- **F2: *Political will to use or increase user fees is lacking.***
- **E1: *Low-cost energy supplies are severely curtailed.***
- **V2: *Environmental improvements are not supported.***

In this scenario, energy costs would increase dramatically by 2001 because of increasing worldwide demand for petroleum. As a result, single-occupancy vehicle (SOV) travel would be curtailed severely; public transit, taxis, gypsy cabs, carpools, vanpools, and high-occupancy vehicles would replace SOVs for a large amount of travel. Nonwork travel would be reduced, and leisure-related travel would be virtually eliminated. Travel by motorcycle, bicycle, pedacycle, and walking would increase, particularly when weather permits.

Congress and the White House could agree to nationalize all industries related to power and energy supply. In addition, petroleum supplies would be rationed; priority would be given to food production; heat, light, and power for essential manufacturing activities; and freight movement to maintain the nation's economy. By this time most local public works departments would have geographic information systems with sufficient data on residential and commercial property parcels, census data, land use, residential and workplace population, and so on to enable the distribution of power for heat and light on the basis of a new federally mandated block grant allocation mechanism. Personal mobility would not be considered a high priority use for petroleum. The supply of funds for highway projects, directly tied to the gas tax, would shrink initially and all available funds would be directed to maintaining an operational system. Limited funds would be allocated to a small number of priority transportation projects, principally rail transportation and public transportation (both intracity and intercity), to support food supply and distribution, the nation's commerce, and work-related travel.

Vehicle and fuels technology could stagnate if long-range research and development (R&D) were cut in both the public and private sectors. Some attempts would be made to implement technologies currently under development, but efforts would stall if success were not immediate; no silver bullets would be available to ameliorate reduced petroleum supplies. Consumers would face long lines to buy federally rationed gas and would begin to demand smaller, lighter cars in order to reduce fuel costs. Safety could be impaired if growing use of smaller cars resulted in increased accident and injury severity.

Carpooling for work trips would increase dramatically, particularly in the suburbs where a large stock of vans and sport utility vehicles would be available for such use. In addition, other trips in the suburbs could be arranged and coordinated through informal neighborhood transportation associations for a wide range of trip types, including shopping, school, and recreation; the coordination could be aided by a cellular telephone boom so that some form of shared transit or paratransit might spring up in virtually every suburban housing development. Entrepreneurs with cell phones and home computers could become major transportation brokers in many suburban communities.

The residential real estate market would become less desirable at locations that rely on low-cost travel. New construction would be limited to infilling at selected locations. Demand for locations that have been developed as (or have evolved into) mixed-use centers would increase. Many people would try to reduce housing costs in the face of rising transportation costs, and families would consolidate residences as many young people returned to their parents' homes. Many housing units near suburban business and commercial locations could be cut up to serve larger numbers of tenants, despite zoning regulations intended to limit residential densities. Growing numbers of retirees and seniors would be faced with a declining real estate market and would be unable to move closer to essential services. There would be increasing pressure in many communities to amend zoning restrictions to permit multiple-tenant housing and mixed-use development.

Telecommunications and information technologies would be used extensively as more workers than ever expected would work at home at least part-time or at telework centers closer to home. Some defense R&D funding would be focused on technologies that substitute, or reduce demand, for petroleum products. Alternative energy sources would be put on line as quickly as possible, and new applications would be adopted as soon as they were feasible.

The public, concerned primarily with economic well-being, would put a lower priority on environmental protection; consequently, some environmental regulations would be rescinded temporarily with the promise of reinstating them in the future. Nevertheless, there would be the possibility of an overall reduction in emissions because of the reduction in automobile-based travel.

Scenario 2: Significant Downturn for the Nation

Although Scenario 2 also paints a bleak picture of the future, one of the driving forces continues in a positive direction: low-cost energy is available.

- **HT2:** ***Highway technology stagnates.***
- **F2:** ***Political will to use or increase user fees is lacking.***
- **E2:** ***Low-cost energy is available.***
- **V2:** ***Environmental improvements are not supported.***

This scenario would be characterized by growing income disparity, continuing suburbanization, deterioration of public services in the central cities, and more widespread utilization of communications and information technologies that would enable more people to work at home. Despite the deterioration in the nation's infrastructure, in this scenario the public would become more and more unwilling to pay to improve or restore it as all levels of government continued to downsize. Relations among the three levels of government would become strained, and the competition for limited funds would intensify.

The tax base in many inner cities and some suburban communities would erode, while the cost of providing essential public services would continue to skyrocket. Representative government at the local level would become more contentious as the competition for services escalated. Location would become the key to economic survival for many cities; the advantages of being a port or a major freight crossroads or a Sunbelt city with a strong economic base would be enhanced in this period. On the other hand, cities that had entered the era experiencing an economic downturn would find it difficult to reverse the trend. Because infrastructure deterioration is exacerbated by cold weather, cities in the North and Northeast would experience more problems than most.

Low-cost energy would remain available because the president and Congress would remain committed to maintaining low-cost energy supplies. Although freight costs would remain low because of low fuel costs and improvements in intelligent transportation systems (ITS) for commercial vehicle operations (CVO), security would become a growing problem in freight transportation—bonded local delivery service would be as much a necessity as a service. Low-cost energy would support continued growth in personal motor vehicle travel; consumers would continue to seek heavier, more powerful, and more accessorized cars. Transit would be almost entirely privatized. Jitneys, shared taxis, and community transportation associations would flourish. Meanwhile, large numbers of low-income workers would have to reverse commute to work in the suburbs.

For individuals who could afford it, a growing array of private services and products would be marketed for individual or group consumption, sparking the growth of new industries. Safety and security would become commonplace concerns; private security companies would develop rapidly as public police protection became more and more limited to major crimes. Walled enclaves with private security services would grow in number for those who could afford them.

Large numbers of aging baby boomers would be affected by these developments, especially those on fixed incomes who would be more directly hurt by deteriorating infrastructure. Despite low fuel prices, other travel-related costs would mount, and heightened security concerns would reduce the willingness to travel for many.

Scenario 3: The Present Reflected into the Future

Scenario 3 presents a future somewhat like the present: technology is widely used, low-cost energy is readily available, and environmental improvements are supported. However, political willingness to use or increase user fees for transportation improvements is lacking.

- **T1:** ***Highway technology is widely used; innovation flourishes.***
- **F2:** ***Political will to use or increase user fees is lacking.***
- **E2:** ***Low-cost energy is available.***
- **V1:** ***Environmental improvements are supported.***

Because the public is convinced that highways deteriorate too quickly, it would demand greater improvement in the existing highway system based on longer-lasting pavements, better road design, and lighter and "road-friendly" vehicles. At the same time, there would be growing public recognition of the importance of environmental issues and the external costs of highway transportation, which would translate into widespread support for technological innovation in highway and transportation systems. Alternative ways would be sought to make more efficient use of existing facilities.

The Federal Highway Administration and the state highway administrations would initiate a whole set of highway projects based on public/private partnerships; the private sector would agree to finance the development, implementation, operations, and maintenance of the infrastructure needed for ITS in several large metropolitan areas. Initially, revenues from ITS/CVO, automated traffic enforcement, private toll roads and bridges, and passenger car ITS operations would generate sufficient revenues to offset the costs and provide an adequate return on private investment. The willingness of the private sector to invest substantially in ITS infrastructure, the need to rehabilitate much of the nation's public works infrastructure, and new federal program incentives would result in new metropolitan-based authorities established to manage the operation and maintenance of all local public infrastructure components (including highways, airports, mass transit, water transportation facilities, sanitation, sewage, water supply, and natural resources facilities). The major beneficiaries would be metropolitan areas with healthy economic potential; the major losers would be communities with poor to low economic potential, heavy location liabilities, and widespread infrastructure rehabilitation needs.

The continuing availability of low-cost energy would yield higher trip generation, less incentive for using public transit, increased tourism and other optional travel, and greater suburbanization. Although passenger and freight travel demand would grow, travel during peak periods would be reduced as more and more people would adopt alternative work schedules through a combination of telecommuting, working at home, working at telework centers, and using staggered work hours designed to better accommodate two-worker family schedules. ITS technology would support a reduction in peak-period congestion.

Despite the public support for environmental issues, a growing conflict would develop in many locations that experienced high growth in the past decade or two as they face the need to choose between rehabilitating aging road infrastructure (something that older, more mature locations have already dealt with) and adding road capacity to serve the burgeoning suburban locations.

Scenario 4: Improvements in Many Areas

Scenario 4 is the most favorable scenario: all driving forces are positive.

- **T1:** ***Highway technology is widely used; innovation flourishes.***
- **F1:** ***Public and political willingness exists to use and increase user fees.***
- **E2:** ***Low-cost energy supplies are available.***
- **V1:** ***Environmental improvements are supported.***

Clear and undisputed evidence of substantial increases in environmental warming would show that in the absence of major changes in lifestyles or other measures to reduce environmental pollution, the coastal land mass would be such a high-risk area that the insurance industry would restructure (greatly reduce) catastrophe insurance on all property within 50 mi of the Atlantic and Pacific coasts, effectively putting a hold on real estate speculation there. Although many landowners would experience considerable hardship, states would begin to strengthen regulations aimed at reversing global warming. Since petroleum-based transportation is recognized as a major contributor to global warming, it would be targeted for improvements to reduce its impacts.

This situation, in conjunction with publicly supported strict environmental regulation and practices, would stimulate breakthroughs leading to an inexpensive, renewable, nonpolluting energy source and a range of innovative technologies that use it. Consequently, the cost of energy would be at a historic low. At the same time, accounting for the societal costs of highway transportation could become the norm for further improvements in the nation's transportation system; states would begin to adopt "green accounting" to meet public demand for greater accountability and would adopt user fees based on potential damage to the environment (not exactly a carbon tax, but close to it).

Because petroleum-based fuels would fall out of use, the traditional sources of revenue for highway agencies would no longer be available. In conjunction with other cost saving and deficit reduction measures, Congress would adopt a unified budget requiring that expenditures be allocated according to need. There would be public support for public transportation, and financing would be made available for transit and other systems to provide access to education, employment, social services, recreation, and other amenities.

COMMON THEMES

Several common themes emerged from discussions about the four alternative scenarios. These themes stem from an effort to sum up the issues that appear to be important to the nation's highway system regardless of the details of the future scenario. The committee sought these common themes because of its initial view that certain issues are likely to be important to the highway system in the year 2020 regardless of the shape the future takes. The committee recognizes that others employing a similar process aimed at the same topic might develop different scenarios and a different list of common themes. However, it believes that many of the themes would resemble the ones found here. The common themes provided guidance to the committee as it prepared its research recommendations. Other bases for the recommendations included the committee's many discussions, its review of the commissioned reports as well as other reports and papers, and the collective knowledge and judgment of its members. The common themes are discussed briefly in the following.

Importance of Communications and Information Technologies

As communications and information technologies continue to change, they are having a growing effect on how businesses locate, manufacture, and ship products and how individuals make decisions about traveling to work, shopping, and conducting personal business. More information is needed about how these rapidly changing technologies affect location and travel decisions and about potential changes in the technologies.

Links Between Highway System and Demographics

With the completion of the Interstate highway system, the nation's network of federal-aid highways is unlikely to change considerably in the next 25 years through a program of new construction although capacity could be added to the existing network, for example, by adding lanes in some locations. However, the driving population is both growing and aging, highway use is increasing, and patterns of highway use for people and goods transportation are changing. As a result, transportation investment decisions at the state, regional, and metropolitan levels are becoming very complicated; although more information about demographics, travel demand, and changes in travel demand is needed, there is considerable uncertainty about how it can be provided quickly and accurately in an affordable manner.

Changing Government Roles

Government roles in transportation continue to change, and the outcome is uncertain. Some current changes will probably continue: for example, more downsizing and devolution, more responsibility for investment decisions at state and local levels, and more use of public/private partnerships, privatization, and other types of innovative financing to supplement traditional public financing of transportation facilities.

Importance of Foreign Petroleum Dependence and Environmental Regulations

Both U. S. dependence on foreign petroleum supplies and environmental regulations are likely triggers for change that could affect the highway system, but highway agencies have no control over either. Nevertheless, in the event that regulations change or petroleum supplies are disrupted, state and local highway agencies, as they have in the past, will look to the Federal Highway Administration for information and guidance.

Volatility of Service and Retail Sectors

Widespread changes are taking place in the service and retail sectors, greatly affecting service provision, goods shipment, and highway travel demand. Additional changes appear likely, both in how these sectors operate and in the travel they generate, requiring the attention of transportation planners and decision makers.

Design of Highway Infrastructure and Vehicles

Changes in vehicle design and construction—affecting the size and weight of automobiles as well as the materials used in them—need to be monitored and evaluated for their effect on highway safety and the environment. Changes in vehicle design, size, and materials should be factored into the design of highways and roadside features such as traffic barriers and crash cushions; supports for signs, luminaries, and utilities; and drainage structures in the right of way. The designer's task becomes more complex as the variability in vehicle size and weight increases.

Changes in Vehicle Fuels and Power Systems

Although technological advances in alternative vehicle fuels and power systems have not affected the dominance of internal combustion engines for the vast majority of highway vehicles, such advances will continue to be made and may be accelerated for environmental or energy policy reasons.

Since changes in vehicle fuels and engine technology can affect vehicle performance, highway financing, and air quality, they need to monitored.

Road Pricing, Travel Demand, Air Quality, and the Economy

While there is considerable evidence that congestion pricing—charging motorists who wish to drive in targeted corridors during peak travel periods—can reduce the amount of peak-period travel in these corridors, little is known about how congestion pricing affects aggregate travel demand, air quality, and the economy of a region. Metropolitan and regional decision makers will need much more information about these issues, as well as the institutional issues involved, before demand management through congestion pricing can be implemented on a widespread basis.

REFERENCES

Abbreviation

FHWA Federal Highway Administration

Bonnett, T.W., and R.L. Olson. 1994. Four Scenarios of State Government Services and Regulation in the Year 2010. *Planning Review*, July-Aug.

FHWA. 1996. *FHWA Entering the 21st Century*. U.S. Department of Transportation, Jan.

Global Business Network. 1994. *Energy and Transportation Task Force: Scenarios Project*. Working Draft Report. President's Council on Sustainable Development, Oct.

Morrison, J. Ian. 1994. "The Futures Tool Kit," *Across the Board*, January.

Wack, P. 1985. Scenarios: Shooting the Rapids. *Harvard Business Review*, Nov.-Dec.

Appendix A

Future Travel Demand

The Impact of Current and Future Societal Trends

Sandra Rosenbloom
Drachman Institute
University of Arizona

A number of important changes in the way individuals, households, firms, and industries interact will have sometimes profound implications for travel needs and patterns in the United States, as elsewhere in the world. This paper highlights the major societal trends that will affect the nature of American travel demand in general and the demand for highway transportation in particular.

The analyses focus on three classes of trends: those that affect *people* directly, those that most directly affect *the economy,* and those that affect the *physical environment.* It is important to remember that these groupings, while a useful way to make manageable a large number of individual trends, are, ultimately, artificial. Most of the trends that affect the economy also affect individuals and their households, sometimes dramatically. And the way that households organize themselves and the activities of their members have profound implications for both the economy and the physical environment.

Figure A-1 suggests the analyses that this paper covers. The first section of the paper describes *current or emerging patterns* in each of the three groupings that might affect travel behavior and ultimately highway use in the United States. The next major section identifies trends over time in those patterns. The next section considers the direct and indirect transportation impacts of those patterns and trends. The final section of this paper considers the research and development needs that follow logically from the trends and patterns described.

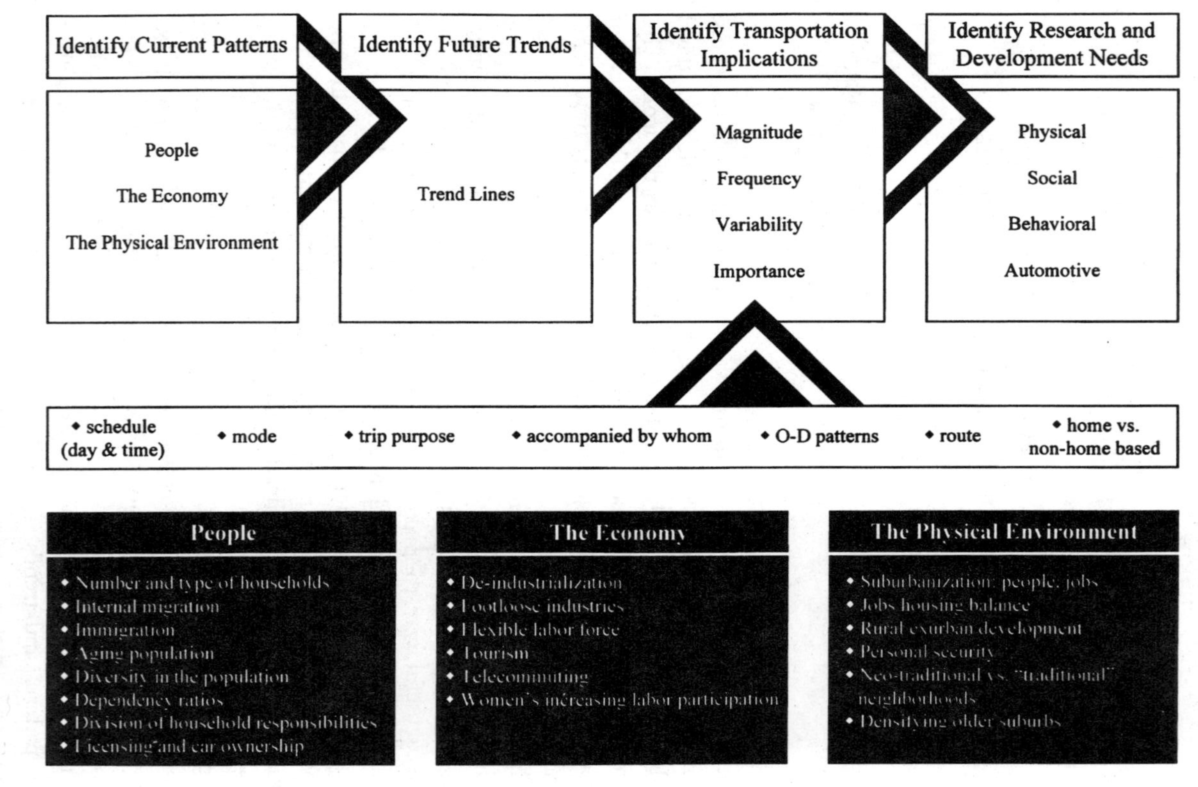
Identify Current Patterns
People
The Economy
The Physical Environment
Identify Future Trends
Trend Lines
Identify Transportation Implications
Magnitude
Frequency
Variability
Importance
Identify Research and Development Needs
Physical
Social
Behavioral
Automotive
◆ schedule (day & time)
◆ mode
◆ trip purpose
◆ accompanied by whom
◆ O-D patterns
◆ route
◆ home vs. non-home based
People
• Number and type of households
• Internal migration
• Immigration
• Aging population
• Diversity in the population
• Dependency ratios
• Division of household responsibilities
• Licensing and car ownership
The Economy
• De-industrialization
• Footloose industries
• Flexible labor force
• Tourism
• Telecommuting
• Women's increasing labor participation
The Physical Environment
• Suburbanization: people, jobs
• Jobs housing balance
• Rural exurban development
• Personal security
• Neo-traditional vs. "traditional" neighborhoods
• Densifying older suburbs

Figure A-1

CURRENT AND EMERGING SOCIETAL PATTERNS

Current People Patterns

The U.S. population has been growing 1.16 percent per year since 1980, reflecting (*a*) a rise in the frequency of childbearing, (*b*) a decrease in death rates, and, most significantly, (*c*) a sustained flow of immigrants from abroad (Hollmann 1992). Large and growing numbers of the U.S. population are from different cultural, racial, or ethnic backgrounds. In 1993 approximately 15 percent of the population was black, 11 percent Hispanic (of any race), 4 percent Asian and Pacific Islander, and just under 1 percent American Indian, Eskimos, or Aleuts (Day 1993).

The U.S. Census predicts that by the turn of the century the white population will account for 84 percent of the total population—down from 87 percent in 1993—while roughly 13 percent will be black, 4 percent Asian or Pacific Islander, and 11 percent Hispanic (of any race). However, by 2050 Hispanics may well compose 23 percent of the population while the white proportion drops to just over half.

Population increase due to birth is actually on the rise in the United States. There has been a substantial increase in the number of births in the United States—the number of annual births rose to 4.2 million in 1990—for the first time in a quarter of a century. Yet just a decade ago demographers predicted a drop in fertility, in large part attributable to their estimate that up to 25 percent of all women born during the so-called baby boom would remain childless (O'Connell 1991). In fact, the rates of childlessness among this group are running only 17 percent—largely because so many of these women simply shifted childbearing to older ages. Most demographers believe that much of the natural increase in the entire population in the last decade was "catch-up" childbearing among baby boomer women in their 30s.

The most important societal trends directly affecting people and their families are the following:

- Aging of the population,
- Changing nature of the American household,
- Family relationships,
- Racial and ethnic diversity of the nation,
- Immigration from abroad,
- Internal migration, and
- Licensing and automobile ownership rates.

Aging Population

American society is aging rapidly; in 1990 more than one-fourth of the entire population was over 60. Indeed, the elderly are the fastest-growing component of the U.S. population; the number of those over 65 grew more than 20 percent between 1980 and 1990. Moreover, in 1990 there were 6.2 million Americans over 85, a number the Census expects to increase by more than 400 percent by 2050. By the first decade of the next century, almost half of all elderly people will be over 75—and almost 5 percent of the entire U.S. population will be over 80. If birth rates continue to drop and migration does not increase, it is entirely possible that more than half of the U.S. population will be over 50 by the middle of the next century (Taeuber 1992).

Like most Americans, in 1990, over three-quarters of all those over 65 lived in metropolitan areas, with almost two-thirds living in the suburbs of those areas. Those elderly people who do live in the central cities of metropolitan areas are more likely to be members of ethnic or racial minorities; they are also more likely to be women living alone and to be poor.

At the same time, more than 8 million Americans lived in nonmetropolitan, or rural, regions in 1990; because younger people have been moving out of nonmetropolitan counties, the actual *concentration* of rural elders has been increasing substantially. Nationally, the rural elderly constitute more than 15 percent of the population in the areas where they live (Rosenbloom 1993) and the oldest old (over 85) are more concentrated in rural areas (Taeuber 1992).

For more than 30 years the residential mobility of older Americans has been dropping. Most elderly people age in the homes in which they lived as younger members of the work force (Taeuber 1992). Between 1986 and 1987, fewer than 2 percent of those over 65 moved far enough to change counties, and fewer than 1 percent moved to another state (Taeuber 1992). In fact, among the elderly who do move, the largest percentage stay within the same region but merely change counties—for example, 60 percent of all moves by those over 65 living in the Northeast in 1986–1987 were to another county within the region.

The suburbanization of the older population has been accompanied by substantial increases in driver licensing. Today most older people are drivers; between 1983 and 1990 the increase in licensing among both older men and women was substantial—not, of course, because older people learned to drive but because younger drivers were aging. In 1992 almost 90 percent of men and 50 percent of women over 70 were licensed drivers; more importantly, almost 100 percent of men and 90 percent of those who will be over 70 in 2012 are currently licensed drivers.

Changing American Households

Today there are more households per capita than ever before, largely because of the dramatic increase in single-person and single-parent families. In addition, the changing American household has new and complicated activity patterns, based in new sets of personal and family responsibilities.

Between 1969 and 1990 the number of American households grew almost 50 percent, while the population grew only 21 percent. Yet the growth in the number of traditional households—that is, a married couple and children—was only 8 percent over that period. The fastest growth rates were among one-person households—an almost 41 percent increase—and single-parent households—a 36 percent increase. As a result, the number of *all* families headed by women alone increased from 11 to 20 percent.

Family Relationships

Those currently of working age have been called the "sandwich generation" because they may have responsibilities to their children and their parents at the same time. This situation arises because many people delayed the birth of their children while their older parents are living longer. A 50-year-old woman could easily have a 15-year-old child and an 85-year-old parent.

A number of studies have shown that "intergenerational linkages" between older people and their younger relatives have been decreasing for years. Between 1962 and 1982 the number of elderly people who saw one of their children at least once a week decreased by 25 percent. There has been an even greater decline in the number of men providing household repairs and women providing domestic help to their aging parents. Most experts see these trends resulting from the substantial increase in the employment of daughters and daughters-in-law as well as the high level of divorce, which weakens intergenerational links (Bumpass 1990).

Caring for children is increasingly a problem for two-worker or single-parent families. A 1993 study concluded that in the last three decades, men have spent more and more time on household activities; between the early 1970s and the mid-1980s they did more "traditional male tasks" such as household repairs and lawn care. Since 1985 men have helped more with what the researchers call "female tasks," such as cooking, cleaning, and laundry (Marini and Shelton 1993). But part of what fueled the closing gap was that the total time that a household spent on domestic responsibilities declined as women entered the labor force.

A 1988 study found that male household efforts have become responsive to the employment status of their wives—men in households with a nonsalaried wife contribute 30 percent less time to household duties than those with a salaried wife. However, the same study found that husbands still carry only a third of the household task load even when the wife has full-time salaried employment (Dutchin-Eglash 1990). In fact, most studies still find that women, even when fully employed outside the home, take on the majority of household responsibilities (Grieco et al. 1988; Van Knippenburg et al. 1990).

A study of panel data from the Michigan Study of Income Dynamics for the years 1979–1987 found that large disparities in the time spent on housework between men and women have continued, even when the wives are also in the full-time labor force. In households in which both spouses had paid employment, men averaged 7 hr/week on housework while women averaged 17 hr; in no case did men conduct, on average, more than 29 percent of all household activities. When children were present, working women averaged 23 hr/week of housework while men still spent 7 hr/week on all household activities (Hersch and Stratton 1994).

Data from the 1987 National Survey of Families and Households indicated that employed women put in an average of 33.8 hr/week in household labor while employed men, by their own report, averaged under 19.1 hr/week of housework. If the paid labor of both sexes was added to household work, men worked 52.6 hr/week while women worked 67.4 hr (Marini and Shelton 1993).

Diversity and Migration from Abroad

The U.S. Census predicts that by the turn of the century, the white population will account for 84 percent of the total population—down from 87 percent in 1993—while roughly 13 percent will be black, 4 percent Asian or Pacific Islander, and 11 percent Hispanic (of any race). However, by 2050 Hispanics may well compose 23 percent of the population while the white proportion will drop to just over half.

Migration is one of the largest causes of both this country's population growth and its growing diversity. Latin America has been the major source of legal immigration to the United States since 1969, the primary country of birth being Mexico. More than 43 percent of the current foreign-born population came from Latin American countries; the bulk of the remainder of legal immigrants has shifted from those of European origin to those from Asia. Today, those born in Asia account for 25 percent of the foreign born, compared with 21 percent from European countries. In fact, in the last half of the 1980s, the total number of Asian immigrants even outnumbered those from Latin America—1.32 million Asian immigrants arrived in the United States compared with 1.02 million Latin Americans (Hollmann 1992).

Internal Migration

Today, as a result of population movement from one part of the country to another, the largest component of the U.S. population (34.3 percent) lives in the South, while the smallest share of the population (under 20 percent) lives in the Northeast region. However, the fastest *growth* in population has been in the Western region, where many states have shown double-digit population increases since 1980: Nevada grew almost 40 percent in the last decade, while Alaska and Arizona also grew over 30 percent. In contrast, two of the states in the Southern region actually declined in population (West Virginia and the District of Columbia), while most of the other Southern states grew less than 7 percent since 1980 (Byerly 1992).

The major internal migrations of this century were the movement of Southern blacks to the northern industrial cities and the movement of large numbers of people to the South and West, particularly California and Florida. In 1920 millions of poor blacks left the rural South looking for better opportunities and jobs. As a result, the distribution of the African-American population changed—from one in which over 90 percent of blacks lived in the rural South to one in which almost half did not (Gober 1993). After the Second World War,

> Migration streams [flowed] from industrial core regions to the southern and western periphery. Industrial cities became major sources of out-migration. The former periphery in the South and West (led by California and Florida) became the cutting edge of economic development and the hot spots on the migration map (Gober 1993).

Migration streams, for example, connected Florida with both New York and New Jersey as retirees moved there after WWII. Over time, however, the migration stream changed: between 1985 and 1990 as many *workers* moved to Florida from these states as did retirees.

Growth in Driver Licensing and Vehicle Ownership

Between 1969 and 1990 the number of licensed drivers went up almost 60 percent, and today licensing is almost universal among drivers of both sexes under 50. Licensing is growing rapidly among the elderly as younger drivers age; among those 30 to 49, almost 96 percent of the men and 90 percent of the women were licensed in 1990. This suggests that (*a*) within 20 years there will be no more than a 5 percentage point difference in the licensing rates of any group of men and women under 70, and (*b*) most *older* drivers—those over 70—will be licensed (compared with less than 60 percent of all women over 65 in 1990).

Vehicle ownership has also grown remarkably. Between 1969 and 1990 the average number of vehicles per household rose from 1.16 to 1.77, while households having two vehicles jumped 117 percent—an annual growth rate of almost 4 percent. At the same time, the number of households without a car fell by a third, so that only 9.2 percent of U.S. households did not have a car. Almost one in five households had three or more cars. As a result of these trends, in 1990 there were actually more cars than licensed drivers (or 1.01 cars per licensed driver) in America.

Of course, vehicle ownership is not distributed evenly through the population. Households headed by older adults are the most likely to be carless; more than 23 percent of all households without a car were headed by a person over 75 years of age, although they constitute only 6 percent of all U.S. households. More than a fifth of one-adult households did not own a car in 1990 (compared with well over half in 1969). Also less likely to own a car were black households, those headed by immigrants, and those without children (Lave and Crepeau 1994).

The number of people without cars is declining; in 1990 fewer than 10 percent of U.S. households did not have at least one car, compared with more than 20 percent in 1969 (Lave and Crepeau 1994). And because carless households were smaller, in 1990 only 6 percent of the entire population lived in one—compared with more than 21 percent in 1969. In short, most people have either a license or access to a car.

Current Economic Patterns

From 1970 to 1990, while the population of the United States grew 1.6 percent annually, employment grew 2.0 percent per year. As a result, in 1992 there were 20 million more people employed than there had been just 10 years earlier (Kutscher 1993). As the work force grew, some industries gained a disproportionate part of the growth; the number of executive, managerial, and technician jobs, more than 50 percent from 1979 to 1990, and professional specialties grew over 40 percent. Jobs in the service sector increased almost 25 percent—but jobs in agriculture and those of operator and laborer actually fell by more than 10 percent (Silvestri 1993).

In the next decade, 25 million new jobs will be added to the U.S. economy; 94 percent of all those nonfarm wage and salary jobs will be in *service-producing* industries. Within the set of service-producing industries, the Labor Department expects just one, the services division, which is currently the largest source of employment in the economy, to account for nearly half of all new jobs created in the next 15 years. More than one-quarter of the projected growth in nonfarm salary employment (7.1 million jobs) will occur in just health and business services (Franklin 1993).

As the nature of the job market has changed, so has the racial and ethnic composition of the entire labor force. In 1980 (the first year in which data were available) minorities of all kinds composed 18.1 percent of the U.S. civilian labor force; in 1992 that figure had grown to 22.2 percent (Kutscher 1993). Blacks increased their share of the labor force by 1 percent annually in those years—roughly the same rate seen during the previous decade—but Asians and others doubled their share of the labor force. In 1980 Hispanics accounted for 6 percent of the labor force, compared with 8 percent in 1992. The growth in both Asian and Hispanic employment is tied to immigration; among Hispanics it is also tied to a higher birth rate, leading to a younger age distribution (Kutscher 1993).

The most important economic trends affecting travel behavior are as follows:

- Women's increasing labor force participation,
- Deindustrialization,
- Flexible labor force, and
- Working at home and telecommuting.

Women's Labor Force Participation

Aggregate growth rates hide the differences between the sexes; from 1970 to 1990 the participation rate of American women increased over 14 percent, but it dropped almost 4 percent for men (Godbout 1993). The participation rate of women 35 to 44 has grown most rapidly; in 1992 more than three-fourths of women in that age group were in the paid labor force

(Kutscher 1993). As a result, almost 60 percent of all women have paid employment, and women now account for close to half of the labor force; in 1992 women made up 46 percent of the total civilian work force (Kutscher 1993)—compared with 38 percent in 1970 (Godbout 1993).

A more striking fact than the increasing number of employed women is the growth in the number of *married* women who work outside the home. In 1990 almost 60 percent of all married women were employed, in contrast to 1960, when less than a third of married women were in the paid labor force (Lagaila 1992). The aggregate figures also hide the dramatic increase in the labor force participation of *women with children*. In 1986 more than 61 percent of married women with children under 18 worked outside the home, compared with only 27 percent in 1960.

Aggregate data also obscure the even more substantial increase in the labor force involvement of married women with *very young* children. In 1960 only 18 percent of married women with children under 6 were in the paid labor force; the comparable numbers were 30 percent in 1970 and 33 percent in 1976. Today, almost 60 percent of married women with young children have salaried employment (while almost 75 percent of married women with children ages 6 to 17 are in the paid work force) (Lagaila 1992).

Moreover, many of the employed women with children under 6 years old had very young children. In 1990 over half of women 18 to 44 who had given birth in the previous year were employed, compared with 30 percent in 1976 (Lagaila 1992). In fact, in 1990 almost half of all mothers of babies under 6 months were in the paid labor force—1 in 12 employed women had an infant (NCJW 1993). A 1990 Department of Labor study found that over 44 percent of all women return to work before their babies are 6 months old, more than two-thirds of those on a full-time basis (O'Connell 1989).

Industrial Restructuring

One of the most striking economic factors of the past three decades has been the significant change in the *sectoral composition* of the labor force—that is, changes in the industries and occupations in which most workers are employed. There has been a major sectoral shift *from* production and agriculture *to* service industries: from work in factories or farms or mines to jobs in, for example, retail sales, public administration, private households, banking, or communications. In the United States the total number of service-sector jobs grew 73 percent from 1970 to 1990 while those in manufacturing grew only 2 percent; jobs in agriculture actually fell 6 percent. As a result, in 1990, there were almost 85 million jobs in the service sector in the United States, or 72 percent of total civilian employment (Godbout 1993).

The United States has seen substantial widening of wage differentials as real wages fell 1.28 percent for *low-skilled* workers between 1980 and 1989. This drop is linked to the drop in demand for low-skilled workers even as the number of such workers increased (largely because of migration from abroad). As a result, today the United States leads most of the world in the incidence of low-paying jobs (OECD 1994).

In 1990 roughly two-thirds of all service-sector jobs were white-collar jobs; of those, however, only 43 percent were highly paid "knowledge workers"—or managers, executives, and a variety of professionals ranging from scientists to lawyers. The remaining white-collar workers had far less lofty jobs, as noted in the *Harvard Business Review:*

> At the lower end of the pyramid in services is an enormous support staff—fully 57 percent of the white collar sectors work force—that toils on the new assembly line of the information economy. Occupations in this category range from sales workers to secretaries to bank tellers and computer operators. In general their educational records are not particularly impressive, nor are their earning power and career opportunities (Roach 1991).

The rapid changes in the industrial structure of the economy have forced many of those already in the labor force to make drastic changes, what the Conference Board's Dan Lacey calls "downward mobility" (*CQ Reporter* 1992). Few former production workers have a meaningful chance for a smooth transition from declining industries into growing ones, largely because of their lack of education and skills; displaced from production jobs, they will be unable to qualify for any of the better-paying positions (Isserman 1994). As a 1992 report by *Congressional Quarterly* noted,

> In many cases the jobs that have been lost will not come back. That means that today's unemployed will have to look elsewhere for jobs. For many that means changing not only employers but also industries, or moving to other parts of the country. Still others will have to undergo retraining in an effort to move into new occupations altogether (*CQ Reporter* 1992).

Many large groups of American workers fall into the "bottom of the pyramid" because they are poorly educated or have low skill levels; this means that they will be limited to low-paying jobs in the service sector, such as janitors, maids, restaurant workers, or sales clerks. The reasons for the lack of education or skill training are complex, but discrimination is clearly part of the problem. For example, several minority groups achieve lower returns to additional educational training than do whites; that is, they are likely to be paid less than other workers with comparable training or degrees (O'Hare et al. 1991).

Flexible Labor Force

A key component of the service sector is the flexible labor force, which contains roughly one-fourth of all American workers. Associated with the growth of the flexible labor force are

- People working schedules that vary over a short period of time;
- People working multiple schedules, often going from one job to the next;
- People working at widely dispersed job locations in short periods of time; and
- People working at multiple locations in short periods of time.

Today, perhaps 34 million people compose the flexible work force—"contingency workers" who are available to respond to different employers' needs.

One component of the flexible labor force is people in temporary employment; temporary help employed by the business sector added 1 million jobs between 1980 and 1992 (Kutscher 1993). Estimates are that between 1982 and 1993, temporary employment increased almost 250 percent, while total employment grew only 20 percent. Temporary employment services place 1.4 million temps each day—three times as many as they did just a decade ago. Forty percent of the companies who are frequent users report that they use temps as replacements for full-time workers (Cook 1994). In fact, Manpower, Inc., with more than one-half million workers, has a larger work force than General Motors or IBM (Castro 1993).

Another component of the flexible work force is those working variable work schedules; in 1991 more than 15 percent of the U.S. work force, or 12 million Americans, had flexible schedules that either allowed or required them to vary the hours they started or stopped work. This was a 25 percent increase in just 6 years (BLS 1992).

A third component of the flexible work force is those with multiple employers at the same time, including so-called contract workers. In 1991 roughly 6 percent of the U.S. labor force, or 1.2 million workers, had more than one job, including contracts with more than one employer; men were more likely (6.4 percent) than women (5.9 percent) to have multiple jobs. A 1994 study of workers with two jobs, or "moonlighters," found that the substantial growth in workers with multiple employers was largely due to increasing rates among women. In 1970 roughly 2 percent of women but 7 percent of men moonlighted; men's rates continued to drop and women's to increase slightly so that by 1994 they converged at 5.9 percent. The study attributed these patterns to several societal trends:

> In many cases, moonlighting reflects the individual's best choice when faced with the need for a flexible work schedule, but in many others it reflects growing economic hardship that threatens the financial stability of families. Moonlighting trends are linked to growing divergence between rich and poor, as well as a general sense that families are working more for less. Multiple-job holding by women has increased in recent years as a result of the increasing percentage of families headed by females, low relative wages, and stagnant male earnings (Kimmel 1995).

Finally, a key component of the flexible work force is those who work fewer than 35 or 40 hr/week. The expansion of the service sector has been coupled with the rapid growth of part-time jobs; the rate of growth of part-time jobs has outpaced that of full-time jobs in almost all developed countries in the past two decades (Godbout 1993). For example, between 1973 and 1990 the annual rate of growth of part-time jobs in the United States was 2.4 percent, compared with 1.8 percent for full-time jobs. A recent Census study estimated that as many as 90 percent of the new jobs created each month are "involuntary" part-time jobs.

Working at Home and in the Car

The 1991 American Housing Survey found that roughly 2.6 percent of the population worked at home, roughly the same percentage as in 1989 and lower than in 1985. However, these figures may blur some important distinctions between those who are self-employed, those working for pay at home, and those taking home work for which they are not additionally paid.

The Bureau of Labor Statistics found that of 20 million people who reported engaging in some work at home as part of their primary job in 1991, only 2 million were actually paid for working at home, and 5.6 million were self-employed. The remaining 12 million nonfarm workers working at home were just "taking some work home from the office" and were not paid specifically for that work. Most of those who did work at home did not do so for much time; over half of those paid for home employment, as well as those self-employed, only worked at home for 8 hr/week (Deming 1994).

A 1993 survey found that there were 7.6 million telecommuters—those working part of their paid week at home; approximately 75 percent were people working in information industries such as programming, accounting, data processing, marketing, planning, and engineering (DOT 1993). These are clearly professions that lend themselves more readily to work at home than do most production jobs. Since these industries may have substantial numbers of employees within the next 30 years, there is a strong possibility that telecommuting may have substantial impact on transportation patterns in the future. The most likely impact on transit use of increasing telecommuting is negative.

At the same time, there have been reports of the growing number of sales people and others who no longer have an office and use their cars as offices as they travel from one site to another. A recent *Wall Street Journal* article estimated that more than 6 million U.S. workers used their cars each day in lieu of offices, a number that some experts think will increase 25 percent by the end of this decade (Shellenbarger 1995).

Current Physical Environmental Patterns

The dimensions of suburban population growth are staggering: while U.S. population rose 56.1 percent in the 40 years since WWII, central cities grew only 49.9 percent. In contrast, the suburban population grew almost 200 percent in the same years. In short, most of the increase in metropolitan population was actually in the suburbs. As a result, even older cities are becoming less dense as low-density suburbs grow up at their periphery (Wachs 1993).

Since 1950 about one-third of the total U.S. population has lived in the central city, but the suburban portions of metropolitan areas increased from 23 percent of total U.S. population in 1950 to 46 percent in 1988 (Forstall and Starsinic 1992). Although central cities grew faster after 1980 than they had after 1970, their growth *rate* was less than half that of the suburbs (Forstall and Starsinic 1992). In fact, the suburbs absorbed almost 76 percent of metropolitan growth during the 1980s (Forstall and Starsinic 1992).

The most important physical trends affecting travel patterns are

- Suburban population and employment growth,
- Concentrated inner cities,

- Changing nature of neighborhoods, and
- Perception of personal safety.

Suburban Growth Patterns

The word "suburb" is no longer automatically a synonym for "*low density.*" The nature of the suburb has been changing with these growth rates: some suburbs resemble traditional downtowns, while others are almost rural. Moreover, some areas have low *average* density but actually contain very high density employment and even residential complexes. And some older suburbs—those developing between 1940 and 1960—are getting old and much denser, through both in-fill and the in-migration of people with larger families (such as immigrants to the United States). As such, how suburban areas affect travel patterns is open to considerable question.

Since 1990 suburban population growth rates have exceeded 2.2 percent a year (FHWA/FTA/LILP 1993). Employment growth in the suburbs has also been substantial. Since 1980 most employment growth has occurred in suburban areas; in 1990 18 of the 40 largest job centers in the United States were located outside traditional downtowns. Moreover, all of the 18 had more jobs than downtown Pittsburgh. For example, in 1990 in the metropolitan New York area, only 5 percent of the work trips from the six most rapidly growing northern New Jersey counties were destined for Manhattan. In Bergen County, New Jersey, an older "bedroom" suburb for New York, employment grew 24 percent (or 80,000) between 1980 and 1990; the new jobs were filled by reverse commuters from New York, by a 33 percent reduction in workers commuting to New York, and by a substantial increase in workers from other suburban counties (TCRP 1995a).

However, as with residential suburbanization, the full implications for travel behavior may depend on whether suburban jobs are in concentrated centers outside the traditional core or more generally dispersed in low-density suburban patterns. Gordon et al. found that in 1980 there were 23 different "centers" in Los Angeles that attracted a substantial density of trips, but 19 centers accounted for only 17 percent of all jobs in the region (Gordon et al. 1989a). However, by 1990 the proportion of jobs in L.A. centers had dropped drastically so that (*a*) only 7 percent of regional employment was located in centers, and (*b*) the number of centers had dropped to 12 (Gordon and Richardson 1994).

Using the Bureau of Economic Analyses REIS data files, Gordon and Richardson found that from 1972 to 1992, substantial employment decentralization occurred almost everywhere in the United States, with the outer suburbs reaching levels of employment previously achieved by inner suburbs (Gordon and Richardson 1995a). Between 1982 and 1987 metropolitan employment growth was the highest in the outer suburbs for all industrial sectors except manufacturing; for example, it exceeded 3 percent in all metropolitan areas (except Milwaukee) and was over 5 percent in five large cities. This outer suburban employment pattern was not a Sunbelt/Rustbelt phenomenon—the highest rate of outer suburban employment growth in the United States was in four disparate communities: Houston, Detroit, Philadelphia, and Los Angeles (Gordon and Richardson 1995b).

Concentration of Central-City Populations

The reverse side of the suburbanization trend is the profound changes occurring in central cities. The mass movement of American families and business to the suburbs has helped to create central cities that differ sharply from those of 50 years ago, in terms of the types of economic activity that take place there and the kinds of families that live there.

Today almost all U.S. neighborhoods characterized by extreme poverty are located in the nation's 100 largest central cities. Moreover, the percentage of the population in central-city census tracts living at "extreme poverty" more than doubled between 1970 and 1990, from 5.2 to 10.7 percent of the central-city population. As the sheer numbers of the poor increase, they are being more concentrated not only within the central city, but within small areas of the central city; the total percentage of the 100 largest central cities' *poor* populations living in extreme poverty tracts increased from 16.5 in 1970 to 28.2 in 1990 percent (Kasarda 1993).

In 1980 2.4 million poor people lived in areas of concentrated poverty in central cities, or 8.9 percent of all poor people in the United States. In 1991 the poverty rate of all families was 17.2 percent in the total of central cities and 7.2 percent in the suburbs. More than 26 percent of central-city families with children were considered poverty households, compared with 11.9 percent of those in the suburbs; the ratio of poverty households to total households was 2.5 times as large in the central cities as in the suburbs (Quigley 1994).

The concentration of the poor has two major effects on an urban economy: the concentration of low-income households increases the per-capita cost of public service provision, and the pressure to provide these services creates substantial budgetary pressures on local governments that have a disproportionate share of the responsibility for service provision. Thus, they are forced to raise taxes, which in turn accelerates the flight of higher-income households and employment to suburban jurisdictions (Quigley 1994).

At the same time, the nature of the employment base in central cities frustrates attempts to decrease poverty by matching central-city residents to central-city jobs. Most central cities experienced *absolute* job growth but those new jobs are very different from those traditionally found in the central business district (CBD)—there are few manufacturing or production jobs and many high-skill information processing and professional jobs (Elwood 1986; Kasarda 1988).

Changing American Neighborhood

One of the signal features of suburbanization is low-density neighborhoods designed to separate homes from one another and from any type of business or commercial activity. As a result, people must rely on cars to meet even their smallest needs; many trips that could be neighborhood-based in denser communities have now become so long that they can only be served by a vehicle.

There has long been a debate over how much the way we structure communities creates the need for a car, particularly for nonwork trips. In part to respond to the declining ability to use public transit or to walk in our current neighborhoods, neotraditional urban design advocates a return to more traditional, higher-density, mixed-use neighborhoods. In such neighborhoods transit and walking are viable options and required drives are shorter (Handy 1995).

Most of these calls for new communities are based, at least in part, on research that shows that denser communities in the United States and around the world have lower car use and higher transit use. Unfortunately, we do not understand which attributes of those denser communities are linked to decreased automobile use or increased walking or cycling, or if it is possible to manipulate or develop certain kinds of urban form or to design neighborhoods in ways that will influence travel behavior in a meaningful fashion.

Perception of Crime

Many Americans are fearful of walking to transit stops, waiting there, or riding on transit vehicles. We do not have good statistics on the actual incidence of transit crime because of the way in which such crimes are reported. In general, an assault or other incident is considered a transit crime only if it happened on a vehicle or in a station; if a crime is committed while a person is walking to or from a bus, or waiting at an ordinary bus stop, the crime is rarely categorized as having anything to do with transit.

However, actual crime statistics are probably not the issue; a number of studies have found that the *perception of crime* is more important than actual crime rates in motivating people's behavior. In many studies women have reported being more fearful on transit vehicles, waiting at stops, or walking to or from a station; a disproportionate share of older women report such concerns. Several large employers or transportation management associations have surveyed workers, asking why they will not or cannot use alternative modes such as the bus; women are two to four times more likely to report fear for their personal safety as a reason for their mode choice (Prevedouros and Schofer 1991; Rosenbloom and Burns 1993).

FUTURE TRENDS: THE WORLD IN 2020

People Patterns

Aging Population

Of all the people trends, the continued growth of the aging component of the population is the most certain. The fastest-growing segment of the older population will be people over 85 in general, and very old women in particular. Among the elderly, women outnumber men by 3 to 2 and are overrepresented among the very old (DOC 1993). In 1991 almost 46 percent of women but only 37 percent of men over 65 were over 75, whereas more than one in four older women were over 80 (compared with fewer than one in five men).

The Census Bureau predicts that by 2010 more than half of all women but only 41 percent of all men will be over 75. Partially because of the age gap between men and women, older women are substantially more likely to be unmarried or to live alone; in 1990 almost 54 percent of women but only 19 percent of men over 65 were widowed or divorced; 16 percent of men but over 42 percent of women over 65 were living alone.

American Household

Although the rate of new household formation will drop, all indications are that the trends that have created so many new units will continue: divorce, woman having babies without husbands, and children living at home, not to marry, but to live alone. The most important issue is the makeup of single-parent households.

Families headed by a woman alone have considerably higher poverty rates than any other type of household—in 1990 over one-third such families were living below the poverty level (Lamsion-White 1992). In fact, the income of families maintained by a woman with no spouse dropped 5 percent in real dollars between 1967 and 1991 (BOC 1992). As a result, families headed by a woman alone constituted a substantial portion of *all* poor families: over 50 percent in 1978 and over 53 percent in 1990. In order to raise themselves just over the poverty line, the average family headed by a woman alone would require an additional $5,661 per year in 1990 dollars (Lamsion-White 1992).

Family Relationships

In the coming three decades most women, especially those with children, will be in the labor force; thus, the dual demands of children and aging parents will become more severe. At the same time, many older people will be women alone, in need of a variety of services.

The ratio of those 50 to 64 to those over 85 has tripled since 1950 and will triple again over the coming 60 years (Gibeau and Anastas 1989). One of the major implications of the growing percentage of the population over 65 is that there will be fewer and fewer younger workers available to pay for, or to directly provide, services for those seniors who increasingly require assistance—including transportation or services that take the place of transportation.

Although the ratio of those over 65 to those 16 to 64 will actually drop—that is, get better—in the next 15 years as the disproportionately large group of baby boomers provide personal and societal support for their parents, in the subsequent two decades the ratio will climb substantially—that is, get worse. This worsening of the dependency ratio is the result of the aging of the baby boomers, which leaves fewer younger people to pay for needed services.

The extent to which husbands and wives will change the balance of household and child care responsibilities over the next 30 years is open to considerable debate. A recent article in the *Wall Street Journal* commented that

> Most couples today are in what sociologists call the transition stage—evolving between "traditional" roles, with women taking sole responsibility for

homemaking, and "egalitarian" roles, with men and women sharing equally the burdens of homemaking and earning money (Shellenbarger 1996).

Migration and Diversity

The extent to which the United States becomes an even more diverse nation by 2020 will be largely the result of immigration policy and laws, the effectiveness of U.S. efforts to keep out people who are not allowed in legally, and the birth rate over time of immigrants (both documented and not). Most analysts believe that the growth of the Hispanic-origin population will be the major element in total population growth; a recent Census report predicted that the Hispanic population will contribute 32 percent of the nation's growth to the end of the century, and almost 40 percent to the year 2010 (Day 1993).

By 2000 there will be 31 million Hispanics; by 2015 the Hispanic population will be double what it was in 1990. In fact, much of the growth predicted for the West and South will come from the 8 million Hispanics that will be added to the population before the end of the century. Almost 81 percent of that number will reside in those two regions, more than half in just Texas and California (Campbell 1994). This trend explains why Texas in 1994 replaced New York as the nation's second most populous state.

Migration is also related to the growing number of U.S. births. Since fertility rates differ by both race and ethnicity as well as country of origin for the foreign born, there are substantial questions about the impact of migration on overall U.S. fertility rates (Bachu 1991). Yet as with native-born women, most variation in fertility rates is the result of demographic factors (education and work force participation) rather than the mother's place of origin or birth, the duration of her stay in the United States, or whether she is naturalized (Bachu 1991). This has led some Census analysts to conclude that the fertility patterns of immigrants "may eventually resemble those of native-born women" (O'Connell 1991).

Internal Migration

Whether people will keep moving from the Northeast and Midwest to the West, or more likely the South, over the next three decades is an open question. In the last decade, while migration to the South has continued, it has slowed to the West, particularly to California. Most internal migration in the past decade has been from the Northeast and Midwest to the South.

So, despite disproportionate Western growth, some analysts believe that the "westward movement of the U.S. population may be coming to an end" as internal migration slows; for example, net internal migration to the West was almost zero in 1988 (Sink 1992). The most conspicuous indicator is that California, the principal recipient of westward migration in the past 40 years, has seen a marked downward trend in migration (Forstall and Starsinic 1992).

Growth in Licensing and Vehicle Ownership

One of the most certain trends is the continuing growth of both licensing and vehicle ownership in absolute terms, although the rate of change will drop. While licensing will probably stabilize at roughly 90 to 94 percent of the entire population, it is possible that vehicle ownership rates will grow beyond one car per driver. Right now a substantial number of families already own more cars than there are household drivers.

Income Trends

For most of the past three decades, there have been significant increases in American household income. From 1967 to 1991 median household money income increased 14 percent in the United States in constant (real) dollars. The Bureau of Labor Statistics has predicted that between 1992 and 2005, per-capita disposable income will increase at an average annual rate of 6.4 percent, reaching $39,000 in 2005—an increase of $21,000 from 1992. When controlled for inflation, this is a 1.5 percent annual growth rate in real disposable per-capita income (Saunders 1993).

However, the first half of the last decade of this century saw fairly weak income growth; overall real median income has grown by only 0.4 percent since the mid-1970s, while employer-provided nonwage benefits fell substantially. Moreover, the income gains of the past 30 years have not been distributed equally across the population. In fact, some groups of Americans—particularly women heading households with children and elderly women living alone—have actually suffered real declines in income in the past decade. For example, the real earnings of young families and those with a high school education or less declined 30 percent from the early 1970s to the late 1980s (Kimmel 1995).

Even among those with increasing real incomes, several subgroups have not seen their incomes grow as rapidly as the total population. For example, in 1987, Hispanics, who are the fastest-growing group in American society, had a median family income of $20,300, two-thirds of the median income of non-Hispanic families. In real dollars the income gap between Hispanics and non-Hispanics grew between 1978 and 1987. Poverty rates for Hispanics increased in those 10 years as well; in 1987, 26 percent of all Hispanic families had incomes below the poverty level compared with 10 percent of non-Hispanic families (Valdivieso and Davis 1988).

In 1987 the median annual income of black families was $20,200, compared with $31,600 for white families. The ratio of black to white income had actually fallen, to 56 percent since 1969, when black family earnings were 61 percent of that of white families. The poverty rate for married black couples was almost twice that of white couples, and higher in 1990 than it was in 1978 (Lagaila 1992).

Economic Trends

Women's Labor Force Participation

Though the rate of increase will drop in the coming decades, the absolute number and percentage of all women employed outside the home will grow rapidly. The Bureau of Labor Statistics estimates that by 2005 almost 64 percent of women but only 74 percent of men will be in the civilian labor force (BLS 1994).

By 2020 all signs are that most women of working age will be the paid labor force—whether or not they have children. Most will be in full-time employment, although many will not work the whole year because of the nature of their jobs (sales, tourism, etc.). A substantial number of women will be in part-time labor, many involuntarily; a significant proportion will also be moonlighting, working two or three jobs to gain a full-time wage.

Participation rates will differ among various subgroups of the population. Black women and men have long participated in the labor force in equal numbers; in the 1950s over half of black women were in the paid labor force compared with less than a third of white women. However, while black women's rates have continued to increase, reaching 58 percent in 1990, rates among men have been dropping: in 1990 only 70 percent of black men over 16 were in the labor force, compared with 77 percent of white men (O'Hare et al. 1991). As a result, the participation rates of white women are roughly comparable to those of black women.

Hispanic women are less likely to be in the labor force than non-Hispanics; in 1992, 52.8 percent of Hispanic women were employed, compared with 57.8 percent of non-Hispanics. Women from Central and South America, however, were more likely to be employed than other Hispanic women and equally as likely as non-Hispanic women; those from Puerto Rico were the least likely to be employed. The differences narrow when controlled for education: 81.2 percent of Hispanic women with a college degree were in the labor force in 1992, compared with 84.5 percent of non-Hispanic women. Among women from Central and South America without a high school degree, participation was actually higher than among comparable non-Hispanic women (Cattan 1993).

Industrial Restructuring

The growth of the service sector will continue, as will the growing income gap; a very large number of workers will be service workers by 2020. At the same time, the *absolute* number of jobs in the goods-producing sector will continue to grow in the next decade—even though that sector's *proportion* of all nonfarm jobs will drop (Franklin 1993). In fact, a Bureau of Labor Statistics scenario projects that manufacturing employment will reverse its absolute downward trend in the next decade, although its proportion of employment will continue to drop. In 1992 employment in manufacturing accounted for only 16.6 percent of the labor force—compared with 33.7 percent in 1950—but it had 2.5 million more jobs than in 1950 (Kutscher 1993).

Retail trade will replace manufacturing as the second largest source of total U.S. employment; it is expected to generate more than 5 million jobs by 2005. Unfortunately, this industry is dominated by part-time, low-skill, "demand little" jobs that offer minimal chance for advancement. Women have traditionally been the dominant participants in this division, accounting for 52 percent of the jobs in 1990, and holding 68 percent of the part-time jobs (DOL 1992).

Overall, the nature of service-sector growth is such that there will be a substantial growth in jobs requiring a bachelor's degree or postsecondary training and in jobs in which most workers have less than a high school education (Silvestri 1993). The nature of these jobs will intensify the growing gap in the wages of low- and high-skill workers. Labor Secretary Robert Reich noted in 1992, "If you're not college educated, you're seeing your real income stagnate or even decline. If you are college educated, your income is growing. The gap between the two is widening" (*CQ Reporter* 1992).

Flexible Labor Force

All indications are that the four components of this part of the labor force will grow: there will be more part-time workers, more people with multiple jobs, more people working for different employers in a short time, and more temporary workers. Estimates are that by the turn of the century, almost half of the work force will be contingency workers.

Working at Home and in the Car

No one knows the full impact of telecommunications and of the growth of industries that are inherently home-based: child and adult day care, for example. In general, fewer people actually engage in these kinds of work arrangements than the media would have us believe, probably less than 2 percent. Moreover, Labor Department statistics suggest that most of those people are, in fact, in low-paying service jobs and not high-tech jobs in which sophisticated communications technology can or does play a role.

At the same time, there is growing interest in the role of telecommunications to substitute for some, if not all, travel from home. So even though a substantial number of people will probably not give up going to the office completely by 2020, a meaningful number may go in for less than 40 hr or fewer than 5 days. This may ultimately affect their residential location choices, as well as those of the industries in which they work.

Physical and Environmental Patterns

Suburban Growth Patterns

Suburban population density in the next 30 years will depend largely on the location of new housing starts, the density of that housing development, and the likelihood that a greater number of people will live in each housing unit (as seen among immigrants to the United States). A 1995 TCRP project report found that most multifamily housing built in the 1980s was built in inner-suburban corridors in low-density configurations—12 to 18 units per acre. However, the study reports that

> much of apartment and condominium construction in the mid-to-late 1990s will be in the new outer suburbs, near emerging edge cities (e.g., Gainesville in northern Virginia and Peachtree City outside of Atlanta) or immediately adjacent to inner-ring edge cities (e.g., Ballston, Virginia, and Atlanta's Buckhead district) (TCRP 1995a).

Predicting the patterns of suburban employment is also difficult. As with residential suburbanization, the full implications for travel behavior may depend on whether suburban jobs are in concentrated centers outside the traditional core or more generally dispersed in low-density suburban patterns. Gordon et al. found that in Los Angeles in 1980 there were 23 centers that attracted a substantial density of trips, but 19 accounted for only 17 percent of all jobs in the region (Gordon et al. 1989a). However, by 1990 the proportion of jobs in L.A. centers had dropped drastically so that only 7 percent of regional employment was located in centers, and the number of centers had dropped to 12 (Gordon and Richardson 1994).

Overall it is difficult to estimate the impact of different suburban trends by 2020 because we do not know (*a*) the extent to which additional suburban employment (or residential) growth will be in concentrated centers versus dispersed locations, (*b*) the rate of population growth in close-in older suburbs versus suburbs on the fringe, (*c*) whether the increased number of people per housing unit within older suburbs (often associated with immigrant populations) will outweigh development in the outer suburbs, and (*d*) the rate of in-fill in existing suburbs. Certainly both growing land use costs and land use regulations will lead to more housing units per acre; the question is where this higher-density development will occur and when.

Concentrations of Central-City Populations

If current trends continue, the inner cities of most metropolitan areas will be highly concentrated pockets of poverty and despair. This will undoubtedly hamper downtown commercial if not business revitalization efforts, because retailers will be worried about attracting people to developments located so near very run-down areas where panhandling, if not street crime, is a problem.

One of the most intractable of societal problems is the growing disparities between the "have" suburbs and the "have-not" central cities. It is not clear how much better—or worse—this can become by 2020.

Our Neighborhoods

Neotraditional designers have offered a new model of urban neighborhood, which, if widely adopted, may have substantial impact on travel patterns. Planners and transportation analysts, however, are hopeful about certain kinds of design changes and less sanguine about others. Critics have serious doubts about several neotraditional design features: narrow streets, changes in the streetscape to impede the car, building at a more human scale, pedestrian amenities. While it is more thoughtful than previous approaches, several analysts think that simple design features will not affect travel in a meaningful way, even if they do create a difference in whether people perceive walking as a realistic activity.

If neighborhoods are designed so that distances are short, major arterials are avoided, the orientation of commercial activities is carefully handled, and there is pedestrian circulation within commercial activities, residents will take more walking trips. Whether they will also take fewer automobile trips or significantly change their entire travel patterns is still open to debate.

TRANSPORTATION IMPLICATIONS OF CURRENT AND FUTURE TRENDS

Introduction

The myriad societal trends discussed above have direct and indirect impacts on travel patterns and ultimately highway demand. They can and do affect a number of major attributes of travel behavior:

- Total number of trips,
- Scheduling of trips,
- Linking individual trips into chains,
- Trip purposes,
- Routing trips,
- Trip origin and destination (O-D) patterns,
- Frequency of trips,
- Variability in trip patterns,
- Trip mode or modes, and
- Substitutability of trips (for one another and for nontravel).

People Trends

Aging Population

The majority of older people will be licensed drivers; in general they will be wealthier than preceding generations of seniors (Ryscavage 1992). And there will be more of them. Today, older people between 65 and 75 make as many as or more trips than slightly younger workers for shopping, and personal business and recreation, traveling as many miles. This strongly suggests that those who retire retain all their "usual" travel patterns—except the work trip—for as long as they can; they shop at the same stores and travel to the same doctors and visit the same friends, largely because they stay in the same neighborhoods where they lived while members of the labor force, and they continue to drive to meet their needs.

Overall, then, we should expect

- An increase in the absolute number of trips,
- An increase in the number of trips per capita, and
- A decrease in transit share and a corresponding increase in automobile trips.

Other transportation implications are not so clear. The greater dependence on the car, even among those over 80, raises major safety questions—will the growth in the elderly driving population threaten to increase accident rates? Data from the National Personal Transportation Survey (NPTS) show that those over 65—who account for roughly 13 percent of the population and 12.4 percent of licensed drivers—account for only 8 percent of all accidents. But when the accident rate of the elderly is calculated by exposure—that is, by miles driven—the result is the well-known U-shaped curve: older and very young drivers have more accidents per mile driven than those in the middle. Moreover, the rate of accidents per exposure increases rapidly with increasing age after 60. In reality, older drivers have lower overall accident rates simply because they drive less—today (NHTSA 1992). But will they in the future?

Whether we will see per-capita increases in accident rates among the elderly will depend on whether the newer generation of older drivers continue to drive less as they age. It is possible that people used to driving will keep doing so without the reduction seen with previous generations. However, even if either all older drivers reduced their driving as they aged, or newer generations of older people had better driving records per mile driven, there will clearly be a growing number of accidents simply because there are so many more older drivers.

In addition, a substantial subset of the elderly are very poor, generally lacking access to cars and even transit services. In 1990 two out of every five poor households in the United States were elderly households. A recent Census study concluded that

> Growth in real income [in the 1980s] was weakest for elderly single householders, especially women, and those elderly households slightly above poverty. The situation was particularly acute for elderly black women liv-

> ing alone—a group whose poverty rate changed very little in the decade (Ryscavage 1992).

To the extent that these seniors are given rides by others, we may see more trips, and more complicated trips, by caregivers. Working women, for example, may link a visit to an elderly mother on the way home from work, first stopping at the grocery store to pick up supplies that the mother needs. Many older people, however, will simply do without.

Changing American Households

The growing number of smaller households has important transportation effects; one person in a small household is likely to make more trips than the same person in a larger household. For example, in 1990 the total daily trip rate of a two-person household was 5.87 trips; this was an average of 2.87 trips per person, compared with 2.94 daily trips for an individual living alone (FHWA 1994)—an 11 percent difference in daily per-capita trip rates.

The difference is even larger in households with only one car. In 1990 a person living alone who had one car made 3.24 person trips per day, which was 19.5 percent higher than the average per-capita trip rate in a two-adult household with only one car (FHWA 1994). And one-person or one-adult households are more likely to have one car per person than are two-adult households. For example, in 1990, almost 80 percent of single-adult households had at least one car, and thus one car per adult, while only 76 percent of two-person households had one or more cars per person (FHWA 1990).

The growing number of households will clearly lead to

- An increase in the number of trips per capita, and
- An increase in automobile trips.

Many of the newer households are headed by a single female parent, who is more likely to be poor. There is still considerable debate about whether such women have longer or shorter commutes; some argue that many single heads of household (as well as poor married women) face a form of spatial entrapment, being forced to take nearby jobs at lower wages. However, such women may also live in the central city but commute to the suburbs for jobs matched to their skill level. As a result they may be traveling longer than workers making more money.

This may explain why in 1990 both married and single urban women with household incomes under $5,000 traveled 33 percent more person miles each day than comparable men and than women in households making $20,000 to $25,000. Those in households making between $5,000 and $10,000 traveled more than 8.5 mi to work; no other group of women traveled that far until they had incomes in excess of $25,000 (Rosenbloom 1995).

Poor central-city residents may also be disproportionately dependent on the private car, that is, beyond what we would expect given their low wages. Probably because many trips from the central city to the suburbs are so difficult to make using public transit, in 1990 urban women with household incomes between $5,000 and $15,000 were more likely to use a car for their work trips than comparable men. Women in households with incomes between $10,000 and $15,000 were more likely to travel to work in a car than men in households making $10,000 *more* (Rosenbloom 1995). Conversely, in 1990 low-income urban women were less likely to use public transit for their work trips than comparable men; more than 8 percent of men but only 5 percent of women in households with incomes between $10,000 and $15,000 used mass transit for their home-to-work commute.

Overall, while some low-income female heads of household will work close to home, we will probably see among these household heads

- Longer work trip commutes,
- A decrease in transit share and a corresponding increase in automobile trips, and
- An increase in the number of reverse commutes (from central city to suburb).

Family Relationships

We are hard-pressed to predict the impact on travel behavior arising because workers are caring for aging parents; some authorities have predicted that women—who, all studies show, carry the overwhelming burden of elder care—may leave the workforce to care for aging parents (thus reducing the number of work trips and increasing the number of shopping, medical, and personal trips). For those women staying in the paid labor force, we will see

- An increase in the number of trips per capita,
- An increase in trip chaining,
- An increase in "serve-passenger" trips,
- A decrease in transit share and a corresponding increase in automobile trips, and
- Increased variability in trip scheduling and O-D patterns.

It is easier to see the impact of *current* child care and domestic responsibilities on the travel behavior of men and women, although the long-term implications are open to debate. Most transportation studies do not show that men are taking on substantial domestic responsibilities in ways we can see in their travel patterns (Hanson and Hanson 1980; Rosenbloom and Raux 1985; Perez-Cerezo 1986; Pickup 1985), although such responsibilities appear clear in the travel patterns of women.

A 1992 survey in Southern California found that employed women were more than twice as likely as employed men to report needing a vehicle to take children to day care and school (Pickup 1985). A 1990 study in four Chicago suburbs found that employed women made *twice* as many trips as comparable employed men for errands, groceries, shopping, and chauffeuring children (Prevedouros and Schofer 1991). An analysis of the 1994 Portland, Oregon, region activity and travel survey found that women heads of household perform more activities, travel more, are more likely to link trips together, and tend to tie more trips into trip chains when they do link trips, than comparable men (Golob and McNally 1996).

The 1994 Portland study found the more that men worked outside the home, the less they engaged in maintenance activities—and the more their spouse did. While the study also found that the more that women worked out of the home the less discretionary travel they engaged in, they found no change in the travel patterns of their male partners. The authors concluded that even among employed women, there are important gender role differences which are reflected in their travel patterns (Golob and McNally 1996).

Analysis of 1990 NPTS data shows that neither marital status nor the presence or age of children in the household had any effect on the travel patterns of husbands, while having substantial impact on the travel patterns of women. Men in two-adult households made 3.2 to 3.3 person trips per day regardless of any other factor; women with small children made 3.5 trips a day (or 9.3 percent more than men with the same responsibilities) while women with children 6 to 15 made an average of 4.0 trips a day (or 21.2 percent more than comparable men) (Rosenbloom 1995).

Given the need to respond to children in an emergency, to chauffeur those children needing rides, and to conduct much of the personal business supporting a household, it seems clear that among salaried women with children we will see

- An increase in the number of trips per capita,
- An increase in trip chaining,

- An increase in "serve-passenger" trips, and
- A decrease in transit share and a corresponding increase in automobile trips.

It is also clear, however, that younger men are slowly taking on more domestic duties when their wives work. However, if household duties do become more evenly distributed, it is possible that both parents then would have a greater need to trip-link and chauffeur children, and use a car to do so.

Diversity and Migration from Abroad

The ethnic and racial makeup of the population may have travel implications. Several major studies have found that cultural and ethnic preferences have important transportation implications (Ho 1994; Bengston et al. 1976; Millar et al. 1986; Wachs 1976). A study of 1990 NPTS data found that blacks and Hispanics were much more likely to use transit and less likely to use a car for all trips than comparable whites. Moreover, black and white women were more likely to use the car than comparable men, while Hispanic women were less likely than comparable men (Rosenbloom 1995). Subsequent analyses found that neither personal nor household income explained these differences.

Whether or not income, or other factors such as residential location in the city, explains the differences between otherwise comparable travelers from different ethnic and racial backgrounds, we may see an increase in the total number of transit trips as immigration continues. It is far more debatable whether we will see an increase in the number of transit trips per capita with a more diverse population.

Migration will also have an impact on travel behavior, both by increasing the diversity of the population and by concentrating similar people in low-density areas. External migration tends to produce concentrations of low-skilled and poorly educated workers, generally in the low-density South and the West, the fastest-growing and now the largest areas of the country. A recent TCRP study found that immigrants to the United States, regardless of income or length of stay, were more likely to use transit than comparable native-born travelers (TCRP 1996).

On the other hand, transit use does drop absolutely the longer immigrants stay in the United States. A recent University of Southern California study of immigrants in Southern California noted that

> this modest transportation behavior is not a permanent characteristic of individual immigrants. Over time, recent arrivals adapt themselves to California society and improve their economic status. Their convergence on the commuting behavior of native borns is one demonstration of the immigrants' assimilation. . . . Transit planners have been the unintended beneficiaries of a liberalized immigration policy and the post-1965 surge in immigration (Myers 1996).

However, their use of transit as a group does not drop to the level among the native born; moreover, new immigrants cross our borders daily. Overall then, with increasing migration we will see

- An increase in the absolute number of transit trips, and
- A smaller increase in average per-capita transit use.

Internal Migration

In general, people have been moving from higher-density industrial cities to lower-density service-oriented cities. Even the goods-producing firms in the South and West have been able to locate in suburban areas to take advantage of cheaper land costs. As a result, the work trip patterns of internal migrants may change remarkably—even if they keep the same occupation in the same kind of firm.

Overall, most internal migrants have moved from higher-density places to lower-density places. As such, they have generally moved from places where it is both possible and relatively easy to use transit, at least for the work trip, to communities where transit services, even for the work trip, are very limited. Unpublished data from the 1991 American Housing Survey show, for instance, that in-migrants of working age who moved to Phoenix were substantially more likely to move further out from the traditional core than those already in the metropolitan statistical area who moved home.

While low-density Southern and Western communities have some (relatively) high-density corridors and concentrated areas where transit services are practical and well-used, the overall movement of population from the Northeast to the South and to the West is likely to lead to

- An increase in the length of work and nonwork trips,
- A decrease in absolute transit ridership, and
- An increase in automobile and drive-alone trips.

Growth in Licensing and Vehicle Ownership

Increased licensing is directly linked to the growth in travel in the last two decades. People with licenses travel substantially more than comparable people without licenses. In 1990 urban women 16 to 64 with a license made 76 percent more person trips and traveled 191 percent more miles than comparable women without licenses, while urban men with licenses made 42 percent more trips and traveled 137 percent more miles than comparable men without. Most important, people with licenses traveled substantially more by car; men in urban areas made 2.8 vehicle trips per day, compared with 0.1 vehicle trips made by those without licenses.

Trip making and the distance covered in each trip grow substantially when licensing is combined with employment. In 1990 employed people with a license commuted roughly 10 mi to work; those without a license commuted 2.5 mi. Since almost 96 percent of people working full time were licensed drivers, it is easy to see why increased licensing, combined with employment, has contributed to a major increase in the miles traveled by Americans.

Overall then, with the almost universal licensing and household vehicle ownership expected by 2020, we will see

- An increase in the length of work and nonwork trips,
- An increase in absolute and per-capita trip making,
- An increase in automobile and drive-alone trips, and
- A decline in absolute and per-capita transit ridership.

Economic Factors

Women's Labor Force Participation

It would be difficult to overestimate the influence of the growing involvement of women in the paid labor force on U.S. travel patterns. The ways in which salaried women balance their domestic and employment responsibilities create far greater and different impacts on the modes they choose, the hours they travel, the routes they take, and they way they organize and combine their out-of-home activities. The 1990 NPTS data show that, at most income categories, working women always make more trips than comparable men, even though men travel more miles than women except at low incomes. Between 1983 and 1990 women increased their per-capita trip making by more than 10 percent—compared with just 6 percent for men—

and increased their person miles traveled by 20 percent—compared with 17 percent for men.

NPTS data show that women in two-adult households with children 6 to 15 years old make 21 percent more person trips than comparable men; those with children under 6 make over 9 percent more trips than comparable men. Single mothers always make more trips than either comparable women or men, probably because they have no one with whom to share the obligations that require travel.

Because they retain multiple responsibilities when they enter the paid labor force, women often link trips together, dropping children at day care on the way to work or going grocery shopping on the way home. For example, more than 40 percent of married mothers with children under 6, but only 30 percent of comparable fathers, linked trips home from work; moreover, those employed mothers made more multiple trip chains. At the same time, single mothers were substantially more likely to link trips than either partnered parent: 47 percent of single mothers with children 6 to 15 linked trips home from work, compared with roughly 36 percent of comparable married women and 27 percent of comparable married men.

In fact, almost all studies have shown that women are substantially more likely to link trips home from work than comparable men, and that women are also more likely to form complex chains, that is, to link many trips together. For example, a 1992 survey in Southern California found that 29 percent of female workers made a stop on the way home compared with 19 percent of men, and that more women made stops on the way to work as well (CTS 1992). More than one-fourth of women workers making a stop *to* work were dropping off children, a detour almost always made five or more days per week (CTS 1992). A 1993 study of Seattle trip diary data found that women were less likely than men to go straight home from work; the authors concluded that "this reflects the role of females in society and the variety of activities they pursue (e.g., shopping, personal business, and recreation) to satisfy personal and household activities" (Hamed and Mannering 1993).

As a result of trip linking, women are more dependent on the car (Strathman and Dueker 1996): an Arizona study found that the more children a woman had, and the younger those children, the more likely she was to drive to work (Rosenbloom and Burns 1993), while the number and age of children had no effect on men's mode choice. Data from the 1990 NPTS show that women in households earning under $30,000 took a higher percentage of all trips in a car than comparable men. The differences were the greatest at the lowest income levels: women in households making under $5,000 annually made 74 percent of all trips in a car, compared with 61 percent of the trips of comparable men.

Overall, the increasing participation of women, particularly with children, leads to

- An increase in absolute and per-capita trip making,
- An increase in aggregate and per-capita distance traveled,
- An increase in trip chaining and the complexity of the chain,
- An increase in automobile and drive-alone trips, and
- A decline in absolute and per-capita transit ridership.

Industrial Restructuring and Flexible Labor Force

It is also hard to overestimate the impact of the deindustrialization of society on travel patterns, although the effects are more diffuse. They are linked to many other trends, including the growth of the flexible labor force and increased female employment. The major transportation, and ultimately transit, impacts of the overall restructuring of national and international industry will arise from different locational decisions made by service firms and industries, growing income disparities, the drop in the number of home-to-work trips, and wide variations in many individuals' work schedules and job locations.

The growth of the entire service sector has important *locational implications*; the growing suburbanization and even exurbanization of jobs are linked closely to the growth of the service sector. Service industries tend to be smaller, and they do not need to be near one another in the way in which goods-producing firms traditionally did. Service firms tend to be widely dispersed within metropolitan and even exurban areas, rather than clustered and concentrated within the core of the city (Frey and Spear 1988; Clark 1988).

The changing industrial base of the country is also substantially altering the commute trip patterns of many workers; they are traveling at different hours, along different routes, and on different days in the week than comparable people two decades earlier. Commuter trips are now spread over a much longer day, with a sizable minority of travelers having variable work schedules or working late at night or early in the morning (Hensher et al. 1994). As a result of both these trends, many low-income workers may be forced to depend on a private vehicle as much as those with much higher income.

TCRP Project H-3, studying ways to attract automobile drivers to transit, undertook an analysis of the relationship between transit and sectoral employment patterns in more than 1,000 U.S. cities in 1990. It was found that employment both in manufacturing and in two of the largest service sectors, wholesale and retail trade, was linked to lower usage of transit. For example, an increase of 10 percent in the share of retail trade employment translated into an 11 percent reduction in transit use. The authors conclude that manufacturing employment now discourages transit ridership because so many facilities are located in suburban and noncentral areas and workers have variable shifts. They found that wholesale and retail trade jobs are associated with less transit ridership because these types of jobs are widely dispersed in neighborhood centers and malls (TCRP 1995b).

Overall, for the whole package of industrial changes occurring in society we will expect

- An increase in absolute and per-capita trip making;
- An increase in the variability of the scheduling and O-D patterns of trips;
- An increase in the number of trips made outside the "traditional" peak periods;
- An increase in aggregate and per-capita distance traveled, for work and nonwork;
- An increase in automobile and drive-alone trips;
- A decline in absolute and per-capita transit ridership;
- An increase in the number of suburban or central-city trips to rural areas.

Working at Home

The growth in the number of people who do not report to an external office may or may not reduce the total number of trips per capita, the amount of vehicular travel, or the total miles traveled, but it certainly will change the nature, routing, and timing of the work trip. Travel may be reduced as the number of days people must commute to work is reduced.

It is, of course, possible that people will move much farther from their work place if they need not travel there daily, using up any mileage saved on the days they do not report to an external job site. It is also possible that those working at home or telecommuting will make longer nonwork trips than they had previously. But whether or not total trip making or mileage increases, all of these patterns will create work trip commutes that defy the traditional definition of the term.

Physical and Environmental Patterns

Suburban Growth Patterns

The average low densities in many suburban areas alone will affect travel patterns. NPTS data show that those living in suburban and rural areas in 1990 traveled 26 percent longer to work (or for work-related activities) than those living in the central city. For nonwork trips, those living in suburban areas traveled 10 percent longer, and those in rural areas 17 percent longer, than central-city residents.

However, with the growth in communications technology and the substantial increase in a variety of service-sector jobs have come dispersed employment locations that can create very nontraditional commute patterns (Barras 1987; Castells 1989). For example, the commutes of suburban and rural residents are twice as likely to be destined for suburban and rural work places as they are for the central city (Gordon et al. 1989b). In fact, in the 35 metropolitan areas that had more than 1 million people, fully 27 percent of all workers crossed a county line to get to work—a 50 percent increase since 1980 (Forstall 1993).

Although the magnitude of the effects may vary with (*a*) the degree of either residential or commercial development within suburban cores and (*b*) the actual location of new developments in what we refer to as the suburbs, it appears clear that increasing suburbanization will lead to

- An increase in absolute and per-capita trip making;
- An increase in aggregate and per-capita distance traveled, for work and nonwork;
- An increase in automobile and drive-alone trips;
- An increase in the variability of O-D patterns;
- A decline in absolute and per-capita transit ridership; and
- An increase in the number of suburban-to-suburban trips.

Central-City Concentrations

Low-skilled inner-city workers are disadvantaged both by the nature of the jobs left in (or coming to) the central city and by the movement of other jobs to the suburbs. As a result, they are often forced to seek the suburban jobs still matched to their skills and become reverse commuters (Straszheim 1980), generally incurring more expensive and longer commutes in both time and distance, with fewer and poorer transit options (UMTA 1972; Greytak 1974; Farkas et al. 1990; Hughes and Madden 1991). There are, of course, a large number of low-skilled jobs remaining in the core of the central city; however, there are not enough to match all the low-skilled workers. Moreover, the skill needed for many low-skilled jobs is higher than that required in the past.

Thus, between 1960 and 1980, the reverse commute, from central city to suburb, grew as much as did the central city-to-central city commute—8.5 percent—to constitute 8 percent of all commuter travel but more than one-fourth of the trips of central-city workers (later data are not available). In 1980 roughly 5 million American workers were traveling from the central city to the suburbs for work, more than twice the 1960 number. Strikingly, 5.6 percent of all those workers used transit for their work trips (compared with 1.6 percent of workers living and working in the suburbs), in spite of the real disadvantages involved.

However, despite their low incomes, and relatively greater dependence on transit, we would expect to see

- An increase in reverse commuting,
- An increase in per-capita work trip length,
- An increase in automobile and drive-alone trips, and
- A decline in absolute and per-capita transit ridership.

Ironically, those people without jobs may not be able to afford to use the transit that is available in their neighborhoods for nonwork trips.

Changing American Neighborhood

As neighborhoods become less dense, we find more and longer trips, most made in the car. If we gain denser neighborhoods, or neotraditional ones, we may see more walking and biking trips. What we probably will not see is walking or bike trips replacing most or any of the trips formerly made by cars. Overall, then, the only pattern that is clear regardless of whether we continue to reduce or increase density is an increase in absolute and per-capita trip making.

Summary

Overall current and future societal trends will increase the

- Per-capita number of trips and thus the absolute number of trips,
- Length of both work and nonwork trips,
- Variability of trip scheduling,
- Variability of trip O-D patterns,
- Number and complexity of linked trips,
- Number of trips made outside the traditional peak periods,
- Number of suburban-to-suburban trips,
- Reverse commuting, and
- Per-capita and absolute automobile use.

RESEARCH AND DEVELOPMENT NEEDS

These trends, and their transportation implications, create a number of technical and scientific questions in four major areas: physical, social, behavioral, and automotive.

Physical

1. What changes are needed in the nation's *highway design and maintenance standards* to deal with the needs of an aging population? Who will be responsible? What will they cost? How effective will they be?

2. What changes are needed to *signing and information systems* to meet the needs of all drivers, particularly aging drivers? Can we improve signing in a way that improves comprehension for all travelers or are some needs in conflict?

3. What changes are needed in *pedestrian facilities and traffic signalization* to encourage walking and make it safe for anyone's neighborhood? Do all pedestrian improvements come at the expense of highway trafffic?

4. What impacts do different types of *neighborhood design* have on internal and external (to the neighborhood) travel patterns? How do high- and low-density American neighborhoods affect individual mode, scheduling, O-D patterns, and trip characteristics? Are the impacts limited to the neighborhood or are they felt regionally?

Social

1. What is the basis of *higher transit use by immigrants* and those from diverse backgrounds? Will it continue?
2. How do *variable employment patterns* (part-time jobs, moonlighting, temp employment) actually affect travel patterns and, ultimately, total demand?
3. How does *caregiving for older parents and in-laws* actually affect individual travel patterns?

Behavioral

1. Which *pricing and other incentives* can change traveler behavior in the face of the trends discussed here—when, for whom, for how long, how often?
2. What is the basis of the *route and scheduling decisions* taken by individual travelers and how do they affect highway demand?

Automotive

1. What *design changes* will facilitate safe travel by older drivers?
2. What impact will *in-car guidance systems* have on safety and mobility?

REFERENCES

Abbreviations

BLS	Bureau of Labor Statistics
BOC	Bureau of the Census
CTS	Commuter Transportation Services
DOC	U.S. Department of Commerce
DOL	U.S. Department of Labor
DOT	U.S. Department of Transportation
FHWA	Federal Highway Administration
FTA	Federal Transit Administration
LILP	Lincoln Institute of Land Policy
NCJW	National Council of Jewish Women
NHTSA	National Highway Traffic Safety Administration
OECD	Organization for Economic Cooperation and Development

TCRP	Transit Cooperative Research Program
TRB	Transportation Research Board
UMTA	Urban Mass Transit Administration

Bachu, A. 1991. Profile of the Foreign-Born Population in the United States. In *Studies in American Fertility*, Current Population Reports, Series P-23-176. U.S. Bureau of the Census.

Barras, R. 1987. Technical Change and the Urban Development Angle. *Urban Studies,* Vol. 24, pp. 5–30.

Bengston, V.I., et al., 1976. *Transportation: The Diverse Aged.* Policy Report 1. National Science Foundation, RANN Research Applications Directorate, Washington, D.C., May.

BLS. 1992. *News*. USDL-92-491. Aug. 4.

BLS. 1994. *Employment and Earnings*. Bulletin 2307.

BOC. 1992. *Money Incomes of Households, Families, and Persons in the United States: 1991.* Current Population Reports, Series P-60-180.

Bumpass, L.L. 1990. What's Happening to the Family? Interactions Between Demographic and Institutional Change. *Demography*, Vol. 27, No. 4, Nov., pp. 483–498.

Byerly, E. 1992. State Population Trends. In *Population Trends in the 1980's*, Current Population Reports, Series P-23-175, U.S. Bureau of the Census.

Campbell, P.R. 1994. *Population Projections for States, by Age, Race, and Sex: 1993 to 2020.* Current Population Reports, Series P-25-111. U.S. Bureau of the Census, March.

Castells, M. 1989. *The Information City: Information, Technology, Economic Restructuring and the Urban and Regional Process.* Blackwell, London.

Castro, J. 1993. Disposable Workers. *Time*, March 29.

Cattan, P. 1993. The Diversity of Hispanics in the U.S. Work Force. *Monthly Labor Review.* November.

Clark, W.A.V. 1988. Urban Restructuring from a Demographic Perspective. *Economic Geography*, Vol. 63, pp. 103–125.

Cook, C. 1994. Temps—The Forgotten Workers. *The Nation*, Jan. 31.

CQ Reporter. 1992. Jobs in the 90's. Vol. 2, No. 8, Feb. 28.

CTS. 1992. *State of the Commute Report, 1992.* Los Angeles, Calif.

Day, J.C. 1993. *Population Projections of the United States, by Age, Sex, Race, and Hispanic Origin: 1993–2050.* Current Population Reports, Series P-25-1104. U.S. Bureau of the Census.

Deming, W.G. 1994. Work at Home: Data from the CPS. *Monthly Labor Review*, Vol. 117, No. 2, Feb.

DOC. 1993. *Sixty-Five Plus in America.* Current Population Reports, Special Studies, Series P-23-178RV. May.

DOL. 1992. *Women Workers: Outlook to 2005.* Report 92-1. Jan.

DOT. 1993 *Transportation Implications of Telecommuting.*

Dutchin-Eglash, S. 1990. *Housework and the Division of Household Labor, 1987–88.* Master's thesis. University of Wisconsin, Madison.

Elwood, D.T. 1986. Spatial Mismatch Hypothesis: Are There Teenage Jobs Missing in the Ghetto? In *The Black Youth Employment Crisis* (Richard Freeman and Harry Holzer, eds.), Chicago University Press.

Farkas, Z.A., A. Odunmbaku, and M. Ayele. 1990. *Low-Wage Labor and Access to Suburban Jobs.* Urban Mass Transportation Administration; Center for Transportation Studies, Morgan State University, Baltimore, Md. Jan.

FHWA. 1990. *1990 NPTS Databook*, Vol. I.

FHWA. 1994. *Urban Travel Patterns. 1990 NPTS.* Office of Highway Information Management, U.S. Department of Transportation, June.

FHWA/FTA/LILP. 1993. *Metropolitan America in Transition: Implications for Land Use and Transportation Planning.* Searching for Solutions—A Policy Discussion Series, No. 10.

Forstall, R.L. 1993. Going to Town. *American Demographics*, Vol. 15, No. 5, May.

Forstall, R.L., and D.E. Starsinic. 1992. Metropolitan and County Population Trends. In *Population Trends in the 1980's,* Current Population Reports, Series P-23-175. U.S. Bureau of the Census.

Franklin, J.C. 1993. Industry Output and Employment. *Monthly Labor Review*, Nov.

Frey, W., and A. Spear. 1988. *Regional and Metropolitan Growth and Decline in the U.S.* Russell Sage, New York.

Gibeau, J.L., and J.W. Anastas. 1989. Breadwinners and Caregivers: Interviews with Working Women. *Journal of Gerontological Social Work*, Vol. 14, Nos. 1-2.

Gober, P. 1993. Americans on the Move. *Population Bulletin*, Vol. 48, No. 3, Nov.

Godbout, T.M. 1993. Employment Change and Sectoral Distribution in 10 Countries, 1970–90. *Monthly Labor Review*, Oct.

Golob, T.F., and M.G. McNally. 1996. A Model of Household Interactions in Activity Participation and the Derived Demand for Travel. Paper presented at the 75th Annual Meeting of the Transportation Research Board, Washington, D.C.

Gordon, P., and H.W. Richardson. 1994. *Beyond Polycentricity: The Dispersed Metropolis, Los Angeles, 1970–1990.* Working draft. Lusk Center Research Institute, University of Southern California, Jan.

Gordon, P., and H.W. Richardson. 1995a. World Cities in North America: Structural Change and Future Challenges. Presented at Pre-Habitat II Tokyo Conference, Sept.

Gordon, P., and H.W. Richardson. 1995b. *Employment Decentralization in U.S. Metro Areas: Is Los Angeles the Outlier or the Norm?* Working paper. Lusk Center Research Institute, University of Southern California, March.

Gordon, P., H.W. Richardson, and G. Giuliano. 1989a. *Travel Trends in Non-CBD Activity Centers*. Report CA-11-0032. Federal Transit Administration.

Gordon, P., A. Kumar, and H. Richardson. 1989b. Congestion, Changing Metropolitan Structure, and City Size in the United States. *International Regional Science Review*, Vol. 12, No. 2, pp. 45–56.

Greytak, D. 1974. The Journey to Work: Racial Differentials and City Size. *Traffic Quarterly*, Vol. 28, No. 2, April, pp. 241–256.

Grieco, M., L. Pickup, and R. Whipp. 1988. *Gender, Transport, and Employment*. Gower, Aldershot, United Kingdom.

Hamed, M.M., and F.L. Mannering. 1993. Modeling Travelers' Postwork Activity Involvement: Toward a New Methodology. *Transportation Science*, Vol. 27, No. 4, Nov.

Handy, S. 1995. Understanding the Link Between Urban Form and Travel Behavior. Paper given at the 74th Annual Meeting of the Transportation Research Board, Pre-print No. 95-0691, p. 1.

Hanson, S., and P. Hanson. 1980. The Impact of Women's Employment on Household Travel Patterns: A Swedish Example. In *Women's Travel Issues* (S. Rosenbloom, ed.), Washington, D.C.

Hensher, D.A., H.C. Battellino, and A. Mackay. 1994. On the Road to Flexible Working Times. *Policy*, Vol. 10, No. 4, pp. 22–27.

Hersch, J., and L.S. Stratton. 1994. Housework, Wages, and the Division of Housework Time for Employed Spouses. *AEA Papers and Proceedings*, Vol. 84, No. 2, May, pp. 120–125.

Ho, A. 1994. Understanding Asian Commuters in Southern California: Implications for Rideshare Marketing. 1994 Annual Meeting of the Transportation Research Board, Washington, D.C.

Hollmann, F.W. 1992. National Population Trends. In *Population Trends in the 1980's*, Current Population reports, Series P-23-175. U.S. Bureau of the Census.

Hughes, M.A., and J. F. Madden. 1991. Residential Segregation and the Economic Status of Black Workers: New Evidence for an Old Debate. *Journal of Urban Economics*, Vol. 29, Jan., pp. 28–49.

Isserman, A.W. 1994. State Economic Development Policy and Practice in the United States: A Survey Article. *International Regional Science Review*, Vol. 16, Nos. 1 & 2.

Kasarda, J.D. 1988. Population and Employment Changes in the United States: Past, Present, and Future. In *Special Report 220: A Look Ahead: Year 2020.* TRB, National Research Council, Washington, D.C.

Kasarda, J.D. 1993. Inner City Concentrated Poverty and Neighborhood Distress, 1970–1990. *Housing Policy Debate,* Vol. 4, No. 3.

Kimmel, J. 1995. Moonlighting in the United States. In *Employment Research,* Upjohn Foundation.

Kutscher, R.E. 1993. Historical Trends, 1950–92, and Current Uncertainties. *Monthly Labor Review,* Nov.

Lagaila, T. 1992. *Households, Families, and Children: A Thirty Year Perspective,* Current Population Reports, Series P-23-181. U.S. Bureau of the Census.

Lamsion-White, L. 1992. *Income, Poverty, and Wealth in the United States: A Chart Book.* Current Population Reports, Consumer Income, Series P-60-179. U.S. Department of Commerce.

Lave, C., and R. Crepeau. 1994. Travel by Households Without Vehicles. In *1990 NPTS: Travel Mode Special Reports.* FHWA, U.S. Department of Transportation, Dec.

Marini, M.M., and B.A. Shelton. 1993. Measuring Household Work: Recent Experience in the United States. *Social Science Research,* Vol. 22, pp. 361–382.

Millar, M., R. Morrison, and A. Vyas. 1986. *Minority and Poor Households: Patterns of Travel and Transportation Fuel Use.* Argonne National Laboratory, Ill.

Myers, D. 1996. Changes over Time in Transportation Mode for the Journey to Work: Aging and Immigration Effects. Prepared for Conference on Decennial Census Data for Transportation Planning, TRB, National Research Council.

NCJW. 1993. *The Experience of Childbearing Women in the Workplace: The Impact of Family Friendly Policies and Practices.* Women's Bureau, U.S. Department of Labor, Feb.

NHTSA. 1992. Crash Data and Rates for Age-Sex Groups of Drivers, 1990. In *Research Note* (E.C. Cerelli), U.S. Department of Transportation, May.

O'Connell, M. 1989. Maternity Leave Arrangements, 1961–1985. Presented at Annual Meeting of the American Statistical Association, Washington, D.C.

O'Connell, M. 1991. Late Expectations: Childbearing Patterns of American Women for the 1990's. In *Studies in American Fertility,* Current Population Reports, Series P-23-176. U.S. Bureau of the Census.

OECD. 1994. *OECD Jobs Study: Facts, Analysis, Strategies.* Paris, France.

O'Hare, W.P., K.M. Pollard, T.L. Mann, and M.K. Kent. 1991. African Americans in the 1990's. *Population Bulletin,* Vol. 46, No. 1, July.

Perez-Cerezo, J. 1986. *Women Commuting to Suburban Employment Sites: An Activity-Based Approach to the Implications of TSM Plans*. Institute of Transportation Studies, University of California, Berkeley.

Pickup, L. 1985. Women's Travel Needs in a Period of Rising Female Employment. In *Transportation and Mobility in an Era of Transition* (G. Jansen et al., eds.), Elsevier, Amsterdam, North Holland.

Prevedouros, P., and J.L. Schofer. 1991. Trip Characteristics and Travel Patterns of Suburban Residents. In *Transportation Research Record 1328,* TRB, National Research Council, Washington, D.C., 1991, pp. 49–57.

Quigley, J.M. 1994. New Directions in Urban Policy. *Housing Policy Debate,* Vol. 5, No. 1.

Roach, S.S. 1991. Services Under Siege—The Restructuring Imperative. *Harvard Business Review*, Sept.–Oct., pp. 82–91.

Rosenbloom, S. 1993. *The Impact of the Americans with Disabilities Act Transportation Requirements on Older Americans.* American Association of Retired Persons, April.

Rosenbloom, S. 1995. Travel by Women. In *Demographic Special Reports: 1990 NPTS.* FHWA, U.S. Department of Transportation, Feb.

Rosenbloom, S., and E.K. Burns. 1993. Gender Differences in Commuter Travel in Tucson. In *Transportation Research Record 1404,* TRB, National Research Council, Washington, D.C., 1993, pp. 82–90.

Rosenbloom, S., and C. Raux. 1985. Employment, Child Care and Travel Behavior: France, The Netherlands, and the United States. In *Behavioral Research for Transport Policy,* VNU Science Press, Utrecht, The Netherlands.

Ryscavage, P. 1992. Trends in Income and Wealth of the Elderly in the 1980's. In U.S. Census, Current Population Reports, Series P60-183, *Studies in the Distribution of Income,* Government Printing Office.

Saunders, N.C. 1993. The U.S. Economy: Framework for BLS Projections. *Monthly Labor Review*, Nov.

Shellenbarger, S. 1995. Work and Family Go Mobile and Wreck Your Sense of Balance. *Wall Street Journal*, p. B1.

Shellenbarger, S. 1996. More Men Move Past Incompetence Defense to Share Housework. *Wall Street Journal*, Feb. 21, p. B4.

Silvestri, G.T. 1993. Occupational Employment: Wide Variations in Growth. *Monthly Letter Review,* Nov.

Sink, L. 1992. Trends in Internal Migration in the United States. In *Population Trends in the 1980's*, Current Population Reports, Series P-23-175. U.S. Bureau of the Census.

Straszheim, M. 1980. Discrimination and the Spatial Characteristics of the Urban Labor Market. *Journal of Urban Economics*, Vol. 7, pp. 114–140.

Strathman, J.G., and K.J. Dueker. 1996. Transit Service, Parking Charges and Mode Choice for the Journey to Work: Analysis of the 1990 NPTS. Paper presented at the 75th Annual Meeting of the Transportation Research Board, Washington, D.C.

Taeuber, C. 1992. *Sixty-Five Plus in America.* Current Population Reports, Special Studies, Series P-23-178. U.S. Bureau of the Census, Aug.

TCRP. 1995a. *Improving Transit Connections for Enhanced Suburban Mobility.* Project B-6. Urbitran Associates. Aug.

TCRP. 1995b. *Policy Options to Attract Auto Users to Public Transportation,* Vol. I. Project H-3, preliminary draft final report. Portland State University (Ore.), Dec.

TCRP. 1996. *Transit Markets of the Future: The Challenge of Change.* Project H-4B, draft report. March.

UMTA. 1972. *Job Accessibility for the Unemployed: An Analysis of Public Transport in Chicago.* Mayor's Committee for Economic and Cultural Development, March.

Valdivieso, R., and C. Davis. 1988. U.S. Hispanics: Challenging Issues for the 1990s. *Population Trends and Public Policy,* No. 17, Dec.

Van Knippenberg, C., M. Kockelkoren, and N. Korsten. 1990. Travel Behavior of Non-Traditional Households. In *Developments in Dynamic and Activity-Based Approaches to Travel Analysis* (P. Jones, ed.). Gower, Aldershot, United Kingdom.

Wachs, M. 1976. *Transportation for the Elderly: Changing Lifestyles, Changing Needs.* University of California Press, Berkeley.

Wachs, M. 1993. Learning from Los Angeles: Transport, Urban Form, and Air Quality. *Transportation,* Vol. 20, No. 4, pp. 329–354.

Appendix B

Roles and Responsibilities of Government
Financial and Taxation Issues

Lee W. Munnich, Jr.
Hubert H. Humphrey Institute of Public Affairs,
University of Minnesota

The reality of federal fiscal constraint during the workout period for the federal budget deficit is likely to put significant financial constraints on transportation infrastructure investments in the coming decade. The Intermodal Surface Transportation Efficiency Act of 1991 (ISTEA) anticipated the shift of funding and responsibility for a wide range of federal functions to the state and local government levels and to the private sector.

PRIMARY TRENDS AND DRIVING FORCES

Fiscal Trends

Federal, state, and local governments in the United States spent $113.3 billion on transportation in fiscal year 1992. About 60 percent of these expenditures were on highways; the other 40 percent were spent on other modes: air, transit, water, rail, and pipelines. Transportation spending represents just 1 percent of all federal expenditures, compared with 9 percent for states and 8 percent for local governments (BTS 1995).

Significant long-term changes in the financing of U.S. transportation infrastructure are shifting roles and responsibilities in the financing and taxation for highway transportation.

- *Spending on transportation is shifting from the federal level to the state and local levels.* From 1982 to 1992 direct federal expenditures for all modes of transportation grew at a much slower rate, 3.1 percent a year, than did federal transportation grants to state and local governments, which grew by 4.5 percent a year. The growth rate in direct state and local spending on transportation (excluding federal grants) was 7.9 percent a year, significantly higher than the growth rate for federal spending over the decade (Table B-1).

- *Transportation spending growth lags behind all other major spending categories of state and local governments.* Growth in state and local transportation expenditures has not kept up with other areas of spending: education, public safety, environment and housing, general administration, and interest on the general debt. From 1982 to 1992 transportation spending per capita in constant dollars increased by about 20 percent. In comparison, the overall growth for all state and local spending per capita was 36 percent for the same period. In fact, transportation spending grew more slowly than any other major area of state and local spending (Table B-2).
- *Over the past four decades, real per-capita transportation spending remains unchanged while other areas of state and local spending have doubled, tripled, or quadrupled.* The shifts in spending at the state and local levels are even more dramatic over a longer period. Since 1957 real transportation spending per capita has remained virtually unchanged, dropping slightly from $311 in 1957 to $306 in 1992. Over the same period, total state and local government spending per capita, in constant dollars, has risen from $1,507 to $3,813, an increase of 153 percent. Education spending contributed to a third of the increase; health and welfare spending added another third: the remaining third can be attributed to public safety, environment and hous-

TABLE B-1 Government Transportation Expenditures
($ billions)

Change	1982	1992	Change	Annual (%)
Federal direct expenditures	9.8	13.4	3.6	3.1
Federal grants to state and local	13.8	21.4	7.6	4.5
State & local direct expenditures	36.8	78.5	41.7	7.9
Total	$60.4	$113.3	$52.9	6.5%

Source: BTS 1995.

TABLE B-2 State and Local Government Expenditures per Capita ($ FY1992)

	1982	1992	Change	Percent Change
Education	1,008	1,298	290	28.8
Health and Welfare	640	967	327	51.1
Transportation	**254**	**306**	**52**	**20.5**
Public Safety	206	304	96	46.6
Environment and Housing	242	307	65	26.8
Government Administration	143	197	54	37.8
Interest on General Debt	130	217	87	66.9
Other General Expenditures	181	217	36	19.9
Total	2,804	3,813	1,009	36.0

Source: State of Minnesota 1996.

ing, interest costs, government administration, and other general expenditures (Table B-3).

Costs for health and welfare, and education and public safety (especially prisons), will continue to place heavy demands on state and local governments. As the federal government cuts back funding in all of these areas, shifting more responsibility to the states, the pressure on states to fund these areas will increase. Opposition to tax increases, particularly the property tax but also other general taxes such as the income or sales taxes, remains strong at the state and local levels. Furthermore, state and local officials are reluctant to increase any taxes that might put them at a competitive disadvantage with other state or local governments.

- *Transportation expenditures by U.S. state and local government have not kept pace with personal income growth.* State and local government expenditures as a percentage of personal income have increased significantly, from 12 percent in 1957 to 20 percent in 1992. However, over the same period, state and local spending on transportation as a percentage of personal income has dropped 35 percent, from 2.5 percent to 1.6 percent (Table B-4).
- *Highway and air transportation have increased their shares of federal transportation outlays while other modes—transit, rail, and water transportation—have lost shares.* Highway spending has increased from 44 percent of federal transportation outlays in 1980 to 51 percent in 1994. Air transportation outlays have increased from 17 percent in 1980 to 26 percent in 1994. The com-

TABLE B-3 State and Local Government Expenditures per Capita ($ FY1992)

	1957	1992	Percent Change	Percent of Growth
Education	535	1,298	143	33
Health and Welfare	255	967	279	31
Transportation	**311**	**306**	**−2**	**−0**
Public Safety	104	304	192	9
Environment and Housing	130	307	136	8
Government Administration	80	197	145	5
Interest on General Debt	41	217	425	8
Other General Expenditures	48	217	351	7
Total	1,507	3,813	153	100

Source: *Census of Governments,* Census Bureau, U.S. Department of Commerce; data prepared by Minnesota Office of the Legislative Auditor.

TABLE B-4 State and Local Government Expenditures as Percent of Personal Income

	1957	1992	Percent Change	Percent of Growth
Education	4.3%	6.8%	60%	32%
Health and Welfare	2.0	5.1	151	38
Transportation	**2.5**	**1.6**	**−35**	**−11**
Public Safety	.8	1.6	93	10
Environment and Housing	1.0	1.6	56	7
Government Administration	.6	1.0	62	5
Interest on General Debt	.3	1.1	247	10
Other General Expenditures	.4	1.1	198	9
Total	12.0%	20.1%	67%	100%

Source: *Census of Governments,* Census Bureau, U.S. Department of Commerce; data prepared by Minnesota Office of the Legislative Auditor.

bined shares of transit, rail, and water and marine transportation have dropped from 38 percent in 1980 to 22 percent in 1994 (Table B-5). The competition for public revenue among modes, particularly between highways and transit, can be expected to continue during the next decade as federal, state, and local funds become even more scarce.

TABLE B-5 Modal Shares of Federal Transportation Outlays

	1980	1994
Highways	44.3%	50.7%
Air	17.2	26.2
Transit	15.1	9.9
Water & Marine	12.7	10.0
Rail	9.9	2.2
Unallocated & Pipeline	.8	1.0
Total	100.0%	100.0%

Source: *National Transportation Statistics,* 1996, Bureau of Transportation Statistics, U.S. Department of Transportation, p.174.

• *Highway revenues at the federal and state levels are generated primarily by motor fuel and motor vehicle taxes, while local highway funding is generated primarily through the property tax and general appropriations.* Most of the public-sector revenue for the financing of highways currently comes from indirect user charges in the form of motor fuel and motor vehicle taxes and direct user charges in the form of tolls. In 1993 user charges accounted for 61 percent of all public-sector revenue for highways. Motor fuel taxes, collected primarily by federal and state governments, represent the largest portion of these user charges, 42 percent of total revenues. User fees account for 87 percent of federal revenue and 78 percent of state revenue devoted to highways. Other sources of revenue that fund highways are property taxes and assessments, general fund appropriations, other taxes and fees, investment income and other receipts, and bond issue proceeds. Local governments generate 92 percent of their revenue for highway financing from these other sources (Table B-6).

• *The burden of funding highways is shifting from the federal to the local level. Highway spending is shifting from the state to the local level.* In 1960 the federal government raised 27 percent of revenues for highways, and local governments generated 21 percent; states contributed the other half of the revenue for highways. By 1994 the federal and local funding roles had reversed, with the federal government contributing 19 percent and local governments raising 27 percent of the revenues for highways. The state share increased slightly, from 53 percent in 1960 to 54 percent in 1994. The state portion of highway spending dropped from 66 percent to 62 percent, and the local share increased from 32 percent to 36 percent (Table B-7).

TABLE B-6 Revenue Sources for Public Sector Financing of Highways in 1993
($ Billions)

	Federal	State	Local	Total	Percent
User Charges					
Motor-Fuel Taxes	$13.8	$22.9	$ 0.7	$37.4	42%
Motor-Vehicle Taxes	2.1	10.7	0.4	13.2	15
Tolls	0.0	3.1	0.6	3.7	4
Subtotal	$15.9	$36.7	$ 1.7	$54.3	61%
Other					
Property Taxes and Assessments	$ 0.0	$ 0.0	$ 4.3	$ 4.3	5%
General Fund Appropriations	1.3	2.1	8.5	11.9	13
Other Taxes and Fees	0.2	1.9	1.4	3.5	4
Investment Income and Other Receipts	0.8	2.2	3.9	6.9	8
Bond Issue Proceeds	0.0	4.0	3.7	7.7	9
Subtotal	$ 2.3	$10.2	$21.8	$34.3	39%
Total	$18.2	$46.9	$23.5	$88.5	100%
Percent of Total	20.6%	53.0%		26.6%	100.0%

Source: *1995 Status of the Nation's Surface Transportation System: Condition & Performance,* Report to Congress, U. S. Department of Transportation, p. 79.

• *Maintenance and other noncapital costs related to highways have increased much faster than capital outlays.* Maintenance and other noncapital highway costs for all units of government have increased much faster than capital costs. In 1960 noncapital expenditures were 38 percent of all highway expenditures. By 1993 the noncapital share of highway expenditures had increased to 52 percent. From 1960 to 1993 noncapital expenditures increased 122 percent in constant dollars compared with total highway expenditure growth of 61 percent (DOT 1995).

• *Future projections for highway capital needs far exceed the revenues that are likely to be available through tax revenues.* According to the U.S. Department of Transportation's (DOT's) 1995 report to Congress on the status of the nation's surface transportation system, the average annual costs through 2013 to maintain

TABLE B-7 Government Receipts and Expenditures for Highways
(Share of Total)

	1960	1980	1994
Government Receipts			
Federal	26.7%	24.8%	18.8%
State	52.7	49.5	53.9
Local	20.6	25.7	27.4
Total	100.0%	100.0%	100.0%
Government Expenditures			
Federal	1.9%	2.2%	1.9%
State	66.2	62.1	61.7
Local	31.9	35.8	36.4
Total	100.0%	100.0%	100.0%

Source: *National Transportation Statistics, 1996,* Bureau of Transportation Statistics, U.S. Department of Transportation, p. 29.

overall 1993 conditions and performance for the highway system are estimated at $49.7 billion, without adjusting for inflation. This is more than 70 percent higher than the actual capital outlay for highways in 1993. Improving the highway system according to economic efficiency objectives would require an average annual investment of $65.1 billion, or 126 percent above the 1993 capital outlay for highways. Given fiscal constraints, it is unlikely that either level of funding will be achieved through public revenue sources.

Implications of Fiscal Trends

These long-term fiscal trends show no signs of reversing over the coming decade. As the federal government seeks to work out the budget deficit, discretionary and postponable spending such as that for transportation investments, unless related to immediate safety concerns, is likely to be subject to at least the same level of cuts as other discretionary areas.

Although tax increases of every sort are currently out of fashion, the primary sources of tax revenue for highway spending have been particularly resistant to increases. The U.S. Congress' Office of Technology Assessment (OTA) laid out the argument for increased fuel taxes in its 1991 report to Congress, *Delivering the Goods: Public Works Technologies, Management, and Financing*:

> Ideally, raising the Federal gasoline tax would encourage higher vehicle occupancy and more efficient use of the highway system, help address traffic congestion and air pollution problems, and reduce the need to build new highways. However, politically the Nation does not seem ready to accept fuel tax hikes of the magnitude necessary to make these sorts of impacts (OTA 1991).

One of the major reasons for increased automobile use and the resultant congestion, air pollution, and need to build new highways is the low gas tax. The fixed cost of owning an automobile has increased 31 percent in real terms from 1975 to 1994. During the same 19-year period, the real cost of gas and oil dropped more than 50 percent. In 1975, 35 cents of every dollar spent on owning and operating an automobile was for variable costs: gas, oil, maintenance, and tires. By 1994, the variable cost share of automobile costs had dropped to less than 20 percent (BTS 1996). This dramatic shift from variable to fixed costs has created a powerful incentive for automobile owners to drive more once they have made the investment in owning an automobile.

A recent public opinion poll reported in *Business Week* indicates just how important the automobile has become to Americans. When asked what invention they could live without, 63 percent said they could not live without the automobile, much higher than any other invention listed. Other inventions included the light bulb, 54 percent; the telephone, 42 percent; television, 22 percent; and the personal computer, 8 percent (*Business Week*, 1996).

Even though there are strong arguments for increasing the gas tax, even a small increase can become politically charged. A recent example is the congressional initiative to roll back the 4.3 cent gas tax increase, which was passed in 1993 as a deficit reduction measure.

Similarly, at the local government level, the property tax has become an increasingly unpopular revenue-raising tool and subject to pressures for funding other critical public services such as schools and public safety. Furthermore, the uneven distribution of property tax capacity among local jurisdictions results in varying capacities to fund the capital and maintenance costs of infrastructure. OTA summarized the problem as follows:

> Traditionally, the Nation's 83,000 local governments have financed most capital investment and all of their operating budgets locally, but their customary broad-based taxes, principally on property, no longer produce sufficient revenue to finance essential services (OTA 1991).

The potential for appropriations for transportation investment and operations from other tax revenue sources is even bleaker. While the federal government seeks to pare back spending in all areas to reduce the deficit, state governments will be burdened with finding the fiscal resources to offset at least a portion of the federal cutbacks in the areas of health and welfare, and to continue to fund increasing demands in the areas of education and public safety. At some point federal and state governments may consider increasing income or sales taxes or enacting a new federal consumption tax such as a value-added tax. However, such broad-based tax increases are unlikely until the public and elected officials have experienced sufficient pain from cutbacks. In any case, the likelihood that these general tax dollars will be available for transportation purposes is remote.

Two related trends, downsizing and devolution, will probably continue as strong policy forces in the United States during the next decade. *Downsizing* refers to the reduction in work force in both public and private organizations in order to reduce costs and streamline organizations to make them more efficient. Downsizing has been driven by increased global competition and the productivity improvement potential of new information technologies. *Devolution* refers to the shifting of powers and responsibilities from the federal government to state and local governments and the citizens. Devolution is part of a long-term process of sorting out how powers should be shared among the federal government and the states. The political momentum for devolution has been fueled by public mistrust of government at all levels, but particularly at the federal level, which is perceived as most distant and least relevant by citizens.

The process of devolution has actually been under way for some time in the area of highway transportation. The development of a national road system in the 1930s and the Interstate system in the 1950s represented a significant shift in authority for the nation's roads to the federal level. As the building of the Interstate system has been completed, responsibilities have been shifted back to the states and local governments. ISTEA began the process of formalizing this shift. ISTEA has meant increased flexibility for states in spending federal transportation dollars as the federal share of transportation funding is reduced.

The downsizing of the federal work force combined with the shift of responsibilities and funding from federal to state and local levels will present significant new institutional challenges and opportunities during the coming decade. The partnership that currently exists among federal, state, and local government transportation agencies in planning and executing transportation projects may help in making this transition. However, there will be a need to enhance planning capacity at the state and regional levels to deal with the increased responsibilities for decentralized decision making.

ISTEA moved surface transportation funding closer to a block grant approach, allowing states and metropolitan planning organizations more flexibility in using funding that had been restricted to highways for transit, bicycle, and pedestrian infrastructure investments. The issue of flexibility in funding is being rejoined in the emerging debate over the reauthorization of ISTEA, as highway advocates compete with transit and community advocates on how scarce public dollars should be allocated.

Federal deregulation of transportation policy is likely to continue, allowing more decisions to be made at the state and local levels. However, major political conflicts may arise over unfunded mandates such as the Americans with Disabilities Act. Congress is unlikely to provide funding earmarked to implement provisions of legislation requiring additional expenditures by state and local governments, but at the same time it may be unwilling to remove mandates for which a strong national political constituency exists.

Environmental and social equity issues will emerge as more important concerns as growing authority for transportation decision making is turned over to states. On the one hand, states and local areas may be more sensitive to the issues in their own backyard. On the other hand, the federal government may be in a better position to set standards for all states where a long-term public concern for energy and the environment or societal responsibility to ensure the equitable treatment of people is involved. The allocation of public dollars between highways, which provide the greatest benefit to those with the highest incomes, and public transit, which is more important to lower-income individuals, is just one manifestation of a transportation equity issue. A broader public dialogue will be necessary to understand and address these issues.

In general, state and local governments will have more flexibility in transportation policy and funding decisions during the coming decade, but there will continue to be a public expectation for federal requirements and standards in such areas as safety, interstate commerce, the environment, and, to a certain degree, social equity.

POTENTIAL SOURCES FOR CHANGE

The financing picture for highway transportation may seem bleak, but the environment of fiscal constraint has the potential for stimulating considerable innovation. There is already evidence of innovative strategies under way at the federal, state, and local levels, which may eventually set the stage for a new U.S. infrastructure investment policy. The following are areas of innovation that could have long-term impact on transportation roles and responsibilities.

Pricing

Transportation pricing could be the most important change on the horizon, particularly when the financial constraints on transportation funding become more obvious to citizens (TRB 1994). Electronic road pricing technology has removed the need for toll booths, allowing drivers to be charged electronically without needing to stop or slow down. This technology is being introduced on many existing toll roads. Many states are considering charging tolls as a way of financing new or expanded roads, either to provide a new general source of funding for transportation improvements or to implement congestion pricing as a demand management strategy. Other forms of pricing include mileage-based charges, emissions fees, and parking pricing.

The political and institutional barriers to implementing pricing strategies are great. The Federal Highway Administration's (FHWA's) congestion pricing pilot program had few takers during the 4 years when funding was available (1991–1995), even with the offer of a sizable federal subsidy for implementing a pilot project. Despite these barriers, several U.S. cities are conducting feasibility studies and are considering congestion pricing as part of a long-term transportation strategy. The first application of congestion pricing in the United States is in California's SR-91 corridor, a privately run road in the median of a public highway in Orange County. The initial public response to SR-91, for which prices vary by time of day, appears to have been positive since its opening in December 1995. The operators of SR-91 initially were unable to keep up with the public demand for the transponders required to use the corridor. The California Department of Transportation reports significantly reduced congestion on the original nontolled lanes of SR-91.

Technology

DOT's intelligent transportation system (ITS) program has encouraged states and local areas to develop and deploy information technologies to improve the efficiency of the transportation system. Current technologies can improve traffic flow, decreasing the need for increased highway capacity and reducing accidents. Advanced traveler information systems could give travelers a better sense of the full costs of travel routes and mode and time decisions. As mentioned, electronic toll collection makes direct user charges more feasible and acceptable to the public, thus offering a new potential source of transportation financing. Telecommunications and information technologies are likely to reduce transportation system demand, though the form and level of impact remain unclear.

Privatization

Transportation services may be privatized at a more rapid rate as financial constraints take hold. California's SR-91 is one of four proposed projects that emerged from a solicitation for proposals from private businesses to undertake transportation projects in the state. This model of soliciting private partnerships on transportation projects has been used by transportation departments in 12 states. There are many obstacles to privatization. For example, experience in Minnesota and Washington has shown that the public will not currently support toll roads on existing free roads or when they expect that public dollars may eventually be available to pay for a new or expanded facility. As the public becomes more aware of the fiscal constraints that will prevent major new investments in highway capacity and if the public sees successful privately run transportation facilities, the private transportation investments may become more common.

Partnerships

Partnerships between transportation agencies and private-sector firms as well as other public agencies are becoming much more prevalent. State transportation agencies have sought out private and public partners to address a wide range of transportation policy issues. The federal ITS program has encouraged states to establish productive partnerships with private firms in testing, evaluating, and deploying new information technologies to improve the efficiency of the current transportation system. These partnerships raise new legal and institutional issues for state departments of transportation (state DOTs). In Minnesota, the state attorney general's office has helped the Minnesota Department of Transportation in addressing legal barriers to forming public/private partnerships under the state's ITS program, Guidestar.

Innovative Financing

Innovative financing, encouraged by FHWA, will increase state and local flexibility in financing transportation projects and will allow the leverage of more private dollars. Under the National Highway System Bill, states are allowed to use federal aid for bond principal, interest costs, and insurance costs; private-sector funds and materials can be credited toward the nonfederal match; federal-aid funds may be loaned to nontoll entities; and states have greater flexibility to use advance construction beyond the life of ISTEA. The bill also authorizes state pilot infrastructure banks to attract private capital and leverage increased financing in private transportation. The states create and operate the banks and may use up to 10 percent of their federal-aid money. Funds in the bank could be loaned out for project construction, or they could be used as loan guarantees to lower interest rates and make projects feasible faster. Ten states will be part of this pilot program (University of Minnesota 1995). The Ohio Department of Transportation has adopted innovative financing as a major strategy for financing long-term infrastructure needs.

Quality and Customer Focus

During the 1980s, U.S. businesses went through a transformation sometimes referred to as the quality revolution. The practices developed by U.S. businesses in response to foreign competition are being adopted slowly by government agencies. One aspect of quality that is becoming much more common in state DOTs is an intense focus on serving customers better. With the constraints on public funding of transportation, state DOTs are focusing more on serving market niches in which there are significant benefits to customers, such as better information or time savings.

Global Competition

Global competition will influence the nature of infrastructure investments and may be the major force driving the United States to invest more in infrastructure despite financial constraints. U.S. businesses have achieved significant productivity gains and have built extensive U.S. markets and distribution systems as a result of the public investments in the Interstate system. Europe, Asia, and developing countries have made and are making major investments in transportation infrastructure as economic development strategies. Strategic investments that reduce congestion without contributing to sprawl or improve the movement of goods through enhanced intermodal facilities will be important in ensuring that U.S. industries remain globally competitive.

Political Conflict

Fiscal constraints have led to intense competition for transportation dollars among donor and donee states, urban and rural areas, and highway and transit funding advocates, among others. Though this political conflict adds enormous uncertainty as to how transportation funding issues will be resolved, it may also offer a unique opportunity for public education on the critical choices in the area of transportation policy.

Citizen Involvement

Transportation agencies are experimenting with new ways of engaging citizens in resolving transportation policy issues. For example, the Minnesota Department of Transportation and the Twin Cities Metropolitan Council cosponsored a 5-day citizens jury on congestion pricing conducted by the Jefferson Center and the Humphrey Institute. Ultimately, citizen understanding and support for transportation policy changes will be necessary to resolve transportation policy and funding issues.

Sustainable Communities

An increasing focus on community sustainability will influence the form of future transportation investments. The Claremont Graduate School and the Surface Transportation Policy Project are collaborating with the Humphrey Institute at the University of Minnesota on a study of transportation and information technologies for sustainable communities for the ITS Joint Programs Office and the Minnesota Department of Transportation. The study focuses on how to develop transportation systems that contribute to sustainability in terms of environmental quality and health, land use, and accessibility. Given the increasing international interest in sustainable development, this new paradigm could have an important impact on the future of transportation policy and funding decisions (State and Local Policy Program 1995).

NEEDED RESEARCH

Researchers should focus on the innovations that occur during the period of fiscal constraint. It is too early to tell which will be the most important elements of a new U.S. infrastructure investment policy. These are a few areas of suggested research that may be important in the future policy:

- *How can full-cost pricing be applied in making transportation decisions at the state and local levels?* The economic tools now exist to apply full-cost pricing in making transportation decisions. The question is how to apply these tools in making transportation decisions at the state and local levels. The Minnesota Department of Transportation has set up an office of investment management to support operating divisions in using full-cost pricing techniques in transportation decisions. FHWA should encourage and provide support in the application of full-cost pricing by state DOTs and assist in evaluating the effectiveness of these efforts.

• *What will be the new federal role in transportation given fiscal constraints?* Interstate commerce, global competitiveness, safety, efficiency, environmental protection, social equity, national standards, research and development, and transportation statistics are all important current priorities for the federal government in transportation. Even with fiscal constraints, the federal government will continue to play a leadership role in all of these areas. However, DOT's role may change to more of a guide and provider of information as funding shifts to the state and local levels.

• *How will information and telecommunications technology change the delivery of transportation services in terms of efficiency, service, and demand?* Changes in telecommunications and information technologies have changed and are continuing to change the way in which all public and private organizations do business. ITS has focused on the potential application of many of these technologies to transportation. A broader question is how telecommunications will affect transportation services. Car phones and telecommuting have already had an effect on travel behavior. As emerging telecommunications and information technologies are adopted by a wider population, how will these technologies affect transportation?

• *What will be the impacts of a more privatized and market-based transportation system? On transportation funding? On economic efficiency? On social equity? On roles and responsibilities of public transportation agencies?* Stephen Lockwood has laid out a vision for a very different way of delivering transportation services in the next century. Lockwood envisions the consolidation of highway agencies with major private transportation and technology companies into regional "transcorps" on a multi-jurisdictional basis (University of Minnesota 1996). This privatized model and other future scenarios for the funding and delivery of transportation services should be examined and considered as a possible basis for moving toward a new national transportation infrastructure policy.

• *What infrastructure investments will contribute the most to productivity and global competitiveness?* It is unlikely that the economic efficiency scenario for highway investment suggested in DOT's conditions and performance report will ever be achieved. However, a better understanding of exactly which transportation investments are most important to U.S. global competitiveness could generate public and political support for funding such projects as an economic development strategy.

- *What is the role of transportation policy in addressing urban sprawl?* The relationship of transportation to land use is becoming an increasingly important question in transportation policy. To what extent do transportation policies contribute to or simply respond to urban sprawl? How should transportation investment decisions be combined with regional growth policies to have the greatest effect? A better understanding of these issues will be important in defining and rationalizing future transportation policies.
- *What are the new partnership models in transportation and to what extent will they become institutionalized (public/private, public/public)?* State DOTs are developing innovative partnerships with businesses as well as other government agencies to improve the delivery of transportation services. Some of these partnerships are short-lived, but others may suggest new models of cooperation for the future of transportation. These partnerships should be studied as potential early indicators of how transportation services may be offered in the future.

In his book *Competitiveness of Nations,* Michael Porter gives examples of industries that have thrived in countries because of selected disadvantages rather than comparative advantages (Porter 1990). A shortage of labor may force firms to find new technologies to make more efficient use of labor. A limitation of natural resources encourages industries to conserve and make more efficient use of those resources. Similarly, the fiscal constraints imposed on transportation by the current political and economic environment may encourage the innovations that will define the transportation infrastructure policy of the next century.

ACKNOWLEDGMENTS

Sandra Seemann, a graduate student at the University of Minnesota's Hubert H. Humphrey Institute of Public Affairs, assisted in conducting background research, organizing focus groups, and preparing this paper. The author also wishes to thank those individuals who contributed to this paper by participating in focus groups to identify key issues or providing suggestions and feedback. Thanks to Robert Benke, Cynthia Burbank, Wayne Cox, Gary DeCramer, Randall Halvorson, Thomas Horan, Adeel Lari, Abigail McKenzie, Robert Morgan, Barbara Nelson, Eugene Ofstead, Carl Ohrn, Clarence Shallbetter, Richard Stehr, John Wells, Robert Works, Lyle Wray, and David Van Hattum for their input.

REFERENCES

Abbreviations

BTS Bureau of Transportation Statistics
DOT U.S. Department of Transportation
OTA Office of Technology Assessment

BTS. 1995. *Transportation Statistics Annual Report 1995: The Economic Performance of Transportation.* U.S. Department of Transportation.

BTS. 1996. *National Transportation Statistics 1996.* U.S. Department of Transportation.

Business Week. 1996. Feb. 19.

DOT. 1995. *Status of the Nation's Surface Transportation System: Condition and Performance.* Report to Congress.

OTA. 1991. *Delivering the Goods: Public Works Technologies, Management and Financing.* Report OTA-SET-477. Washington, D.C., April.

Porter, M. 1990. *The Competitive Advantage of Nations.* Free Press, New York.

State and Local Policy Program. 1995. *Transportation and Information Technologies for Sustainable Communities Interim Report.* Claremont Graduate School and the Surface Transportation Policy Project. Humphrey Institute of Public Affairs, University of Minnesota, Minneapolis, Oct.

State of Minnesota. 1996. *Trends in State and Local Government Spending.* Office of the Legislative Auditor, St. Paul, Feb.

TRB. 1994. *Special Report 242: Curbing Gridlock: Peak-Period Fees to Relieve Traffic Congestion.* National Research Council, Washington, D.C.

University of Minnesota Center for Transportation Studies. 1995. *Getting There from Here: Views on the Financing and Future of Transportation.* Summary proceedings from a policy makers' forum on Innovative Transportation Financing, Minneapolis, Nov.

Appendix C

Communication and Information Technologies and Their Effect on Transportation Supply and Demand

Carol A. Zimmerman with Christopher Cluett,
Judith H. Heerwagen, and Cody J. Hostick,
Battelle Memorial Institute

Throughout history, transportation supply and demand has been the result of a complex set of forces. A recent U.S. Department of Transportation (DOT) report identified six types of factors that affect travel patterns: economic, social, demographic, technological, land use and housing, and transportation policy (DOT 1995a). Technology is only one of many potential influences, and its role will undoubtedly vary in relation to the strength of other factors affecting transportation as well as the rate of technological change at any point in time. Any particular field of technology may go through a period of relative stability, then explode into a period of rapid evolution—such as the current explosion in communications and information technology, which is the subject of this paper. Consequently, the impact on transportation will be more profound during these periods of rapid change. Nevertheless, transportation supply and demand will result from the interplay of all factors, and the impact of communications and information technology will be played out in that context.

TRANSPORTATION SUPPLY AND DEMAND: PUTTING TECHNOLOGY IN PERSPECTIVE

There are several logical ways we can think about communications and information technologies as having an effect on transportation supply and demand. They can increase supply by causing additions to the stock of roadway infrastructure or vehicles; they can increase demand for travel in terms of numbers and length of trips, and numbers of vehicles; they can reduce demand by creating alternatives to vehicular travel; or they can leave supply and demand relatively unaffected while altering the travel experience in terms of efficiency, speed, comfort, predictability, and safety. It is more difficult to see how communication and information technologies might lead to reductions in supply, though indirect effects are plausible. For example, reduced demand associated with substantial mode shifts or the restrictions of selected roadway segments to dedicated use could yield reduced supply.

Personal Travel

DOT's assessment of personal travel in the United States has shown that Americans are making more trips to more locations at more times of the day than ever before (DOT 1995a). These travel patterns are a result of the continuing suburbanization of residences and jobs in metropolitan areas, which has given rise to longer work trip commutes and more suburb-to-suburb travel compared with the central-city dominance of earlier eras. At the same time, household and economic structures have changed such that there are more women working outside the home and more single-parent households. In such households, trip chaining is commonly practiced to ensure that household needs can be met. These trends have meant an increasing reliance on automobile travel.

Transportation demand can be measured in several useful ways, such as the number of vehicles, number of vehicle miles traveled (VMT), persons per vehicle, person miles traveled, number of trips, and length of trips. By whatever measure, transportation demand has been rising steadily and dramatically (DOT 1995b). While the number of households has increased, the number of vehicles per household has increased as well. Over the past two decades the number of trips taken has increased three times the rate of population growth. As employers increasingly locate businesses in the suburbs, inner-city residents must travel farther to access those jobs. As the population ages, the elderly are driving more than they used to; there are many more senior citizens than there used to be, as well.

The structure of the U.S. economy and work force composition has also influenced personal travel. The growth of the service-sector economy has led to the need for a "flexible labor force," whose members tend to work variable schedules and hours, at different locations, and for multiple employers. To accommodate this level of flexibility, workers must be able to have continually varying travel patterns, both geographically and temporally. One countervailing trend is the increase in working at home and telecommuting, although the travel implications are not entirely clear.

Recent analysis of the 1990 Nationwide Personal Transportation Survey (Pisarski 1992) indicates that transit's share of all national travel had declined to about 2 percent at the beginning of this decade. Many urban transportation planners and agencies are struggling with rapidly growing levels of traffic congestion coupled with declining air quality; they are seeking ways to encourage car pooling, transit usage, and alternatives to vehicular travel such as light rail, car and van pooling, and telecommuting, while also trying to curb overall increases in travel demand with growth management and land use policies, employee trip reduction policies, congestion pricing, parking policies, and other strategies. The challenge is daunting because people do not appear willing to abandon their cars or the sense of personal freedom that they derive from their cars, and automobile manufacturers are very successfully supplying the marketplace with appealing automotive products.

Although the intelligent transportation system (ITS) program includes elements of demand management and multimodal applications, its primary emphasis is on enhanced efficiency and throughput of our existing transportation infrastructure. This strategy is likely to encourage the supply of new automotive products and to provide continuing support for the demand for personal transportation; many argue that it will induce even higher levels of demand to take advantage of smoother-running systems. The ITS program also incorporates innovative communication and information systems, such as navigation devices and up-to-the-minute information on traffic and roadway conditions. Whether these systems on balance serve to increase travel demand further by making travel easier and more efficient, or reduce it by enhancing the reliability of alternative forms of access, remains to be seen.

Transportation supply is a more difficult concept to deal with than transportation demand in this context. Supply can be operationalized in several ways. For example, it can be thought of in terms of the number of vehicles available to drivers, production of new vehicles, miles of existing roadway, expansion of existing roadway, miles of new roadway construction, availability of various forms of public transportation, or alternatives to vehicular roadways (bicycles, dedicated bikeways, pedestrian pathways). Supply is measured in terms of both vehicles and infrastructure. The history of the development of the automobile in the United States can be characterized in terms of the expansion of supply, including the construction of the Interstate highway system and the expansion of urban roadway systems to accommodate suburbanization and its associated land use developments. The Intermodal Surface Transportation Efficiency Act of 1991 recognizes the current limitations on our nation's ability to continue to expand the supply side of the equation and emphasizes approaches that put the current "supply" of infrastructure to better use while reducing adverse impacts on the environment and enhancing mobility for a wide diversity of citizens.

Transportation supply and demand are, not surprisingly, closely intertwined; changes in one affect the other and vice versa. Communications and information technologies can, in turn, affect both supply and demand. Advanced traveler information systems, a component of the ITS program, are focused on providing pretrip and enroute information on roadway conditions, accidents, route planning, multimodal choices, and the like to travelers. It is simply too early in these programs to know how this information will influence supply and demand for transportation. Some data suggest that information that makes transit trips, for example, more reliable and predictable by providing riders with accurate arrival times at each scheduled stop may encourage increased ridership. Communications between bus drivers, law enforcement, and transit dispatchers, along with live video on board the vehicle, may increase riders' sense of security on public transportation, which could increase demand. In these ways communications and information can enhance demand and indirectly induce more attention to the supply side of public transportation systems.

Commercial Freight Movement

Three trends in the commercial freight supply chain are continuing to emerge that will affect motor carrier supply and demand. First, margins are becoming tighter because of global competition, making the transportation component of product cost more important. Also, industry continues to move to smaller lot production and mass customization. The net result is that the U.S. manufacturing base is becoming more decentralized, and smaller production facilities are increasingly being distributed closer to population centers to reduce transportation costs. This trend of sourcing industrial production closer to population centers will reduce the demand for long-haul, Class 7 and 8 truck carriers and increase the demand for smaller-class truck shipments in and around population centers.

A second trend that will affect freight supply and demand is the move toward nonstore retailing of goods. Kurt Salmon Associates (KSA) estimated that in 1992, 85 percent of all retail sales were through stores and that 15 percent of retail sales were through nonstore retailing. Nonstore retailing includes phone orders and computer-based interactive shopping (KSA no date). For the year 2010 only 45 percent of all retail sales are expected to be through stores, thus 55 percent of sales will be through nonstore retailing. KSA also indicates that an ever-increasing portion of all types of retailing will require next-day delivery. This trend will also drive freight supply and demand away from long-haul carriers and toward air carriers coupled with less-than-truck load or smaller-class truck freight shipments.

A third trend includes the continued growth in truck volumes and weights and the move toward intermodal shipping containers to improve efficiency. Long-haul motor carriers will be moving larger and fewer loads than in the past. When population trends are factored in, overall long-haul traffic will still be higher than it is today.

RELATIONSHIP BETWEEN TRANSPORTATION AND TELECOMMUNICATIONS AND INFORMATION TECHNOLOGIES

Transportation, telecommunications, and information technologies are entwined in a complex dance of mutual reinforcement and stimulation. New technologies in this realm can both substitute for and stimulate travel demand, and travel itself feeds the invention of new technologies to improve the movement of people, services, and goods (Niles 1994; Mokhtarian 1990). Adding to the complexity are higher-level societal and economic forces that drive the development of new communications and information technologies. Continued pressures for competitiveness feed the development of processes that lower costs and add value to products and services. Globalization of markets leads to the creation of telecommunications services that can connect remote workers with centralized services and with one another and for moving materials in more customized forms and in shorter times than ever before. This section highlights some of the important advances that have been made in technologies that may have a bearing on transportation in the future.

Advancements in Telecommunications and Information Technology

A number of technological advancements have contributed to the increasingly complex relationships between transportation and telecommunications and information technology. These include the following:

TABLE C-1 Evolution of Computer Chip Memory

Amount of Memory	Date Introduced	Cost/MIPS($)
16 kilobits	1976	NA
64 kilobits	1979	100,000
256 kilobits	1982	10,000
1 megabit	1984	1,200
4 megabits	1987	800
16 megabits	1991	15
64 megabits	1993	11
256 megabits	1995*	NA
1 gigabit	1998*	NA

* projected
MIPS = millions of instructions per second
Source: *Business Week* 1995a. *Washington Post* 2/14/95; Midwinter 1995.

1. *Computer and telecommunications equipment is increasingly portable, powerful, and affordable because of advances in microelectronics* (*Newsweek* 1995; Gates 1995). These technologies make it possible for people to work or access information and services anywhere, anytime, as long as the telecommunications infrastructure is available. As illustrated in Table C-1, enormous advances in the power, cost, and size of integrated circuits have helped to fuel the proliferation of equipment available for businesses and consumers alike.

As integrated circuits have become more powerful, they have become smaller. Indeed, miniaturization has been identified as one of the most important technological trends of the decade. "Miniaturizing creates enormous opportunities to put things precisely where they're needed," according to Stephen Millett of Battelle (Sandro 1995).

2. *Americans are becoming increasingly comfortable with computers in many facets of their work and personal lives.* As illustrated in Figure C-1, households have begun to spend more of their disposable income on computer equipment and services. The World Wide Web is making the Internet readily available to millions of people. Estimates of Web usage vary enormously, but they all point to a continued phenomenal growth rate (*Time* 1995; Hoffman and Novak 1995). Increasing use of the Internet for internal organizational work (Intranet) is also growing rapidly, and is an important factor in the growth of mobile and remote work relationships.

Despite the growth in use of personal computers (PCs) and on-line and Internet services, not every American is participating. Census data and other

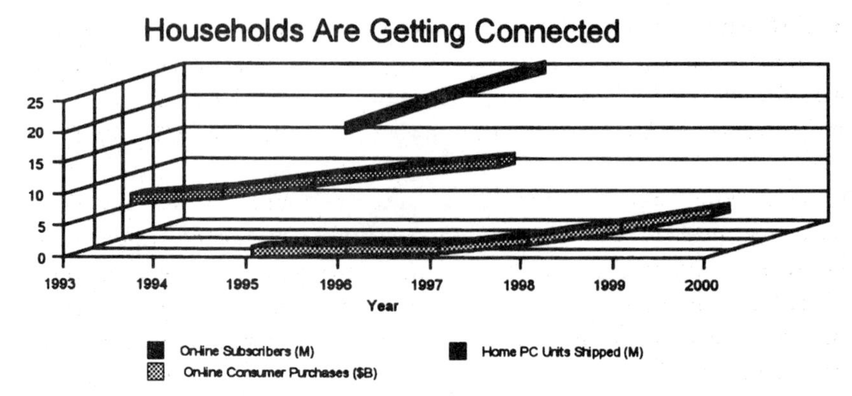

Source: *Business Week,* September 4, 1995, September 25, 1995, November 13, 1995.

FIGURE C-1 Household purchases of computer equipment and services.

surveys document that usage is heavily skewed toward persons who are likely to have higher incomes, have a college education, be males, and be young adults than the population in general (U.S. Census Bureau 1995; *Washington Post* 1995b). If PC and on-line usage follow the normal technology diffusion process, the demographics will undoubtedly change. Whether they will become as commonplace as television sets in the American home remains to be seen.

3. *The merger of computers and telecommunications technologies known as the National Information Infrastructure or Information Superhighway is enabling new kinds of products, services, and relationships* (Gates 1995). This convergence has become possible because of the movement toward digitalization of all types of information—voice, data, and image—which can be transmitted, stored, and processed in a similar manner (Midwinter 1995).

4. *New software technologies, such as intelligent agents, data management, and visualization processes, will enable information to be stored, manipulated, retrieved, altered, and filtered.* All of these are value-adding processes with the potential to create new consumer and work demands and processes (Negroponte 1995; Niles 1994).

5. *Security and privacy concerns associated with the telecommunications technologies, especially the Internet, are likely to continue until secure transmissions are*

available. Although numerous corporations are developing software for secure transmission of credit card data and other private information, the technologies are not yet available for mass use. When these problems are resolved, many expect electronic commerce to expand at a rapid pace (*Time* 1995).

6. *Expansion and restructuring of the telecommunications industry are taking place.* The Federal Telecommunications Act of 1996 marked a significant crossroads in the industry, because it removed many of the barriers that prevented firms in one market from entering another. Many state governments have also been relaxing regulation of the industry within their borders. The competitive floodgates have been opened, and the promise of new products and services and competitive pricing appears to be at hand. Consolidation of the industry has begun (e.g., the Bell Atlantic/Nynex and Disney/ABC mergers). However, further muddling the picture is the fact that many of the firms who are competing in one market are partnering in others. For example, six major cable TV firms are owners of Primestar, a direct-broadcast satellite TV service (*Washington Post* 1996).

The Federal Communications Commission (FCC) has contributed to the growth of wireless, another segment of the telecommunications industry. Rather than give away spectrum licenses as in the past, the FCC has begun to auction portions of the radio spectrum for commercial applications. For example, PCS, the new generation of wireless telephones, yielded the FCC millions in revenue in 1995 as firms large and small vied for licenses in metropolitan markets. As the FCC makes available other parts of the spectrum, industry is proposing new applications, such as crash-avoidance radar and wireless local-area networks in the 47- to 76-GHz range (*Business Week* 1996).

7. *Emerging interface technologies will make it much easier for most people to interact with their computers.* The technologies are more intuitive and visual, with social aids (such as Microsoft's Bob) that help guide the user by providing feedback and tips, or visual navigational aids that use common imagery rather than abstract icons to give a sense of place. For instance, the interface for hotel information may include a street with the hotel shown that, when clicked on, provides room availability, prices, photos, or a filmed walkthrough of the hotel. Gates (1995) notes that interfaces can be programmed to provide real-time video that enables viewers to check out a place, such as a restaurant or store, to see how crowded it is and, thus, whether it is worth making a trip. Voice-activated interface technologies are

an active area of research; the aim of these tools is to enable a person to give the computer (or other appliance) a voice command.

This technology is finding direct applications in the transportation sector with the recent marketing of a voice-command direction-finding navigation system for the private automobile. In-vehicle navigation systems such as this will affect travel behavior in uncertain ways.

8. *Customization will become a dominant theme in the future.* The use of intelligent agents to screen and manipulate data will provide individual users with information tailored to their own preferences and needs. The application of agent technology combined with sophisticated data capture and data analysis processes will enable companies to identify consumer preferences and purchasing habits at the individual level and thus target their marketing and sales efforts much more effectively. This process, called mass customization, enables goods to be both mass produced and customized for individuals. The technology will enable many products to be produced on the spot; for others, it will reduce drastically the time needed to identify a product, order it, find it in the warehouse, pay, and deliver. All of these processes will be connected electronically.

9. *Telepresence is a young technology, but visionaries such as Gates (1995) and Negroponte (1995) see it as an integral part of our lives in the not too distant future.* Telepresence simulates an environment as if it were really being experienced. In its current form, the user generally wears special goggles and gloves that allow him or her to see and manipulate things in a virtual space. The technology is already being used in vehicle simulation and is being developed for remote surgery. Telepresence offers the potential for people in different geographical locations to attend virtual meetings by having their images project into a simulated room and interact as if they were really present.

10. *Personal phone numbers are in the development stage.* This technology would enable a person to have one phone number that goes with him or her everywhere, thus making it easier to communicate (as long as the telecommunications infrastructure is available to place and receive calls).

11. *Further on the horizon, but looming as a substantial force in 21st-century life, is the development of technologies that "think" (Kelly 1994; Brand 1988) and will be able to respond to human behavior.* A recent article in the *Dallas Morn-*

ing News (reprinted in the *Seattle Times*, June 2, 1996) describes these new technologies as "invisible, intimately involved in our lives, pervading our homes, offices, cars, and even clothes." Thinking devices would include in-vehicle monitors that can pick up preferred driving patterns and give advice to the driver, intelligent name cards that exchange information with new acquaintances, and shoes that monitor vital signs. Although these technologies are just in the development stage, the capabilities that they engender point to a new class of computer tools that merge with and react to human behaviors and capabilities, rather than merely expand or enhance them as technologies now do.

Impact of Technological Developments on Transportation Supply and Demand

Impact on Passenger Travel

In a recent examination of the relationship between telecommunications and travel, Niles (1994) develops a conceptual framework that we have found to be compelling. He analyzes the relationship among telecommunications, trip generation, and trip substitution. He contends that trip substitution operates basically at the microlevel of individual transactions, whereas trip generation reflects macrolevel socioeconomic patterns.

Trip elimination occurs when telecommunications processes replace travel by allowing people to communicate from geographically dispersed locations, by allowing information to be sent electronically rather than in physical form, and by allowing people to electronically access documents and data. Examples cited by Mokhtarian (1990) and Niles (1994) include

- Telecommuting by working at home at least part of the week rather than commuting to the office;
- Delivery of entertainment and sporting events to dispersed audiences, primarily at home, as opposed to attending the event in person;
- Use of telecommunications services to gather information on product availability and thus avoid unnecessary trips;
- Researching and purchasing products via electronic commerce;
- Attending classes via distance learning rather than on-site classrooms;
- Accessing telemedicine or other health care services that eliminate trips to health care sites;
- Conducting government services such as field inspections and form processes through voice, data, or video transmissions rather than site visits;

• Working from remote sites in the aftermath of emergencies (such as earthquakes and storms) rather than commuting.

Although these examples point to the ways in which telecommunications can decrease travel demand, the reverse can occur also, with telecommunications serving as a stimulus for travel. Among the stimulant effects cited by Niles (1994) and Mohktarian (1990) are the following:

• Telecommunications processes create new jobs and enhance economic growth, leading to income growth at the societal, organizational, and individual levels with more money available for travel;
• The existing use of telecommunications expands the number and geographic scope of social and economic relationships, which eventually creates new demands to meet in face-to-face settings;
• Telecommunications services lead to expectations for rapid response—and thus commercial organizations may speed up delivery processes, putting more and smaller delivery vehicles on the roads;
• Telecommunications technologies permit people to live at great distances from their places of work, leading to more decentralized residential settlement and to longer trips between residences, work, and services;
• The new, mobile technologies encourage more travel because of the ease of working anywhere, anytime; for sales forces and independent contractors, the efficiency of accessing central data bases may decrease the time per customer and thereby free time to see additional clients; the greatly enhanced access to information about events and people may lead to increased desire to participate and thus stimulate trips that would not otherwise be taken.

Impact on Commercial Freight Movement

Technological developments in communications and information processing have translated into several important applications for improving productivity in the movement of commercial freight, including

• *Automated weigh station, toll collection, credential purchases concepts.* Existing communications and information technology is being adapted to eliminate freight carrier delays due to weigh station inspections, ports of entry, tolls, credential/permit purchases, and the like. Most of the automation concepts use various carrier-mounted sensor devices that collect data for transmission by radio frequency (RF) to receiving stations positioned periodically along freeways. This RF architecture can also be used to enable the receiving stations to automatically "write" via RF signal to carrier-mounted information systems. This automated read-write capability that can be done without the carrier stopping forms the basis for eliminating many of the delays experienced by the freight carrier industry.

- *Automated vehicle location (AVL) and global positioning system (GPS).* AVL is accomplished through an in-vehicle unit that sends location data to a control center via a radio or cellular network link (Brocano 1995). AVL systems rely on GPS navigational data. The GPS constellation of 24 satellites sends navigational data to the GPS receiver within the AVL. The GPS receiver determines vehicle location to 30 to 100 m accuracy using three satellite signals. The resulting radio signal from the vehicle-mounted AVL is fed to mapping software residing at the control center. The mapping software shows vehicle location, direction, and speed and can be zoomed in to show street-level data. Since the control centers can communicate to the motor carriers, the location and speed data can drive dispatch and scheduling systems to optimize truck routes, resulting in more efficient travel as well as maximizing backhaul opportunities to avoid deadheading or traveling empty.
- *RF-generated bills of lading.* Passive RF tags are being developed that are approaching traditional bar coding in terms of cost competitiveness. Passive RF tags have a small chip that can store product data and transmit them to an RF receiver when energized by an electromagnetic signal. They are called "passive" because they have no power source and emit no signal unless energized. RF tags currently cost less than 25 cents each. Once the use of passive RF tags is widespread, vehicle-mounted RF receivers will be able to generate in real-time a bill of lading. The electronic bill of lading data can then be communicated to carrier control centers via AVL communication mechanisms described in the previous section. Real-time inventory data will reduce receiving and inspection costs and eliminate other inefficiencies that exist in freight transport.

Although the emergence of automated weigh station concepts will eliminate motor carrier delays, reducing the cost of motor freight as compared with other modes of transportation, the cost reduction will not be significant enough to appreciably alter motor freight supply and demand. The emergence and widespread use of AVL/GPS and RF tag technology will reduce motor carrier freight transport costs, but since the same technology can be implemented by other modes of transportation, the relative mix of transport modes (and resulting supply and demand) most likely will remain unaffected.

Given that communications and information technology is emerging that can automate data handling and decision making associated with motor freight transport, the next step is in the direction of automated material handling and container designs that leverage this capability. It is anticipated that intermodal containers will be "modularized" to enable loads to be loaded and unloaded and automatically handled in less-than-truckload units. For example, traditional 20- and 40-ft intermodal containers might be replaced with smaller containers that are interlocking and can be handled as one large container, or "burst" to support automated distribution of freight at interfaces between long-haul routes and urban distribution centers. Transport costs will be reduced in this scenario, but it is not anticipated that the supporting technology will greatly alter supply and demand patterns.

Impact of Transportation Demand on Communications Technologies

Transportation demand can also stimulate the development and availability of new communications technologies, which may, in turn, make it easier and more convenient to travel. Mokhtarian (1990) notes that the need to monitor the movement of passengers and cargo has led to the development of navigational devices, tracking and locational equipment, and cellular communications devices. The most often cited example is the tremendous growth in cellular phones over the past 5 years. Traffic congestion is less of an irritant to drivers who can use the time to conduct business.

In addition to the ways telecommunications can stimulate travel demand, a number of powerful vested interests are also operating to keep vehicles on the road. The automobile industry, hotels, airlines, restaurants, shopping centers, sports, tourism—all exist because people travel. There is little incentive on the part of these interests to reduce travel demand; in fact, the new network technologies increasingly are being used by commercial vendors to encourage people to shop, eat, tour, and be entertained, preferably away from their homes.

IMPLICATIONS FOR RESEARCH AND DEVELOPMENT

The explosion in the creation and deployment of communications and information technology may lead to changes in transportation supply and demand, as indicated in the preceding section. In trying to plan for and manage the changes to the transportation system that these technologies engender, the Federal Highway Administration and other transport-related agencies need to direct some of their research and development (R&D) activities toward the subject. This section identifies some potential areas of R&D.

Research on Traveler Information Needs

Only when information is targeted to the expressed needs of potential recipients (individuals or organizations) is it likely to influence their behavior and have the desired effect on transportation demand and supply. We need well-designed assessments of who is accessing the information provided, how they are using that information, whether they interpret the information in the same way that the sender intended it, what they like and dislike about both the medium and the message, and how it is connected to behavioral changes. This evaluative user feedback should be reflected in alterations in the content of the communications and a process of "shaping" the technology to better fit the needs and uses of the recipients. A balanced approach is called for between the effective marketing of information technologies and a highly participative process in which users help design the products in ways that enhance the likelihood that we will see desirable changes in transportation supply and demand.

Testing Theories of the Impacts of Telecommunication on Transportation

Earlier works on transportation and telecommunications (Mokhtarian 1990; Niles 1994) for the most part are derived theoretically rather than empirically. There is a great need for data in this area to either support or refute some of the key relationships they identify. Research focused on further identifying and understanding the relationships between telecommunications and transportation supply and demand would be valuable. A key issue is whether to study these at a macro level or at a more reductionist level where it is possible to see specific linkages better. Maybe a combination of both approaches is needed. Some specific thoughts on potential research in this area include the following:

Use of Communications and Information Technologies in Travel Decision Making

As has been pointed out, the relationships between information technologies and transportation supply and demand are exceedingly complex. Efforts to influence transportation demand through policies that attempt to alter driver behavior patterns through regulatory means or with economic incentives have met with substantial resistance. From a policy standpoint, we simply do not know how to present communication and information technologies to predictably alter transportation supply and demand. In fact, we do not even understand how communication and information technologies affect transportation supply and demand when we are not trying to influence the outcome. A first-order research requirement is to seek some understanding of the key processes. Case studies could be implemented in selected urban areas to evaluate carefully how different households respond to the new information technologies. This might include studies of telecommuters to understand how their travel behavior is different from that of those who do not telecommute, or to examine how driver decision making is affected by the availability of real-time traveler information. Since the aggregate societal effects that we observe reflect the sum of many individual driver decisions, we need to better understand how new communications and information technologies are influencing those decisions. Similar studies could focus on the organizational level to understand how organizational behaviors are altered by the introduction of these information technologies. We need to conduct our R&D on manageable pieces of this complex puzzle and, as we gain new insight into the underlying relationships and processes, we can begin to put the puzzle together and understand the whole.

Alternatively, and perhaps simultaneously, we could more closely examine aggregate effects and not try to disentangle the complexity of the interactions. We could conduct comparative case studies of urban areas of similar size and character in which one is experimenting with various traveler information systems and the other is not. Then we could take periodic measures of indicators of travel demand and supply to see how those change over time. Such indicators might include, for example, changes in travel time in various corridors, in modal choice patterns, in the provision of transportation infrastructure, and in average VMT, number of trips, length of trips, and trip-chaining behavior. If we observe patterns of change in these aggregate measures, then we could work down into a more detailed assessment of how and why these changes are occurring.

Research on Relationship Between Consumer Behavior and Technology

With the ever-increasing means of delivering travel information to consumers, it would be useful to know how the different technologies (e.g., Internet, telephone, kiosks, etc.) influence decisions on transit mode (e.g., take the car or bus) and time of travel. Are people more likely to travel by bus if they are able to get timely information on bus location and arrival times? Are they more likely to carpool if they are given information on coworkers who live near them? Many large corporations are implementing trip reduction programs and may be willing to participate in such studies.

Economic Effects of Emerging Communications and Information Technologies on Freight Transportation Supply and Demand

The development of communications and information technologies has the potential to reduce motor carrier freight costs and, thereby, increase the demand on motor carriers versus other modes of transportation such as rail and air. By quantifying the price points by freight type that drive demand decisions, and by estimating how communications and information technologies would change anticipated freight cost by transportation mode, motor carrier market share can be projected to estimate future impacts on motor carrier demand. Similarly, understanding how these technologies reduce freight inefficiencies can help in determining the effect on supply. For example, if technologies can reduce the need to deadhead or backhaul empty trailers by a certain percentage, an understanding of what portion of trailer movement empty backhauling activity can be used to estimate the potential reduction in such truck miles.

Use of Visualization Tools in Modeling Relationships Between Technology and Travel Behaviors

The complex web of relationships between telecommunications technology and transportation presents a major difficulty for researchers and policy makers. The traditional means of showing relationships with boxes, lines, and arrows can be overwhelming when the relationships include positive and negative effects, feedback loops, and indirect or second-order impacts. Advanced visualization and multimedia technologies could provide important conceptual tools to help researchers and policy makers identify key relationships and communicate them to others. These tools would be especially helpful in developing and assessing alternative scenarios and building hypotheses that could be tested in the field.

We recommend a research program that would join information technology specialists and social scientists in the development and testing of tools that conceptualize and visualize relationships in ways that can be grasped easily. The tools should be interactive and allow for increasing levels of complexity as the user works with the programs. In this way, the tools would be useful for learning purposes and not just as static images to be used in presentations.

Implications for Policy Development and Program Implementation

The title of this paper begs an interesting question. Are we dispassionately interested only in how communications and information technologies might effect transportation supply and demand? Or are we more interested in understanding the underlying relationships in order to exercise public policy to create desired transportation outcomes? Urban areas reflect a dynamic tension between the effect of market forces and contemporary planning on urban form and function. Experience shows that growth management and transportation planning and policy have a limited ability to alter the transportation behaviors of individuals and organizations (Hodge et al. 1995). These efforts raise a number of important policy research questions about what kinds of effects are reasonable to expect and how those effects are experienced by citizens and organizations. We need to understand how communications and information technologies operate synergistically with other policy strategies—that is, better information might facilitate the effectiveness of other policies in altering demand and supply in desirable ways. We need to know how these effects are distributed across the urban landscape—that is, are the impacts experienced equitably by all segments of the population? Who benefits and who loses in terms of these changes? Do the second-order economic outcomes from changes induced in demand or supply benefit some and not others? How are the costs of the communications and information technologies distributed across society? What opportunities are there for public/private partnering in sharing the risks and benefits of deploying these technologies?

REFERENCES

Abbreviations

DOT U.S. Department of Transportation
KSA Kurt Salmon Associates

Brand, S. 1988. *The Media Lab: Inventing the Future at MIT.* Viking Penguin, New York.

Brocano, S.K. 1995. The GPS-RF Connection for AVL. *Communications Magazine*, Nov., pp. 15–18.

Business Week. 1995a. The Technology Paradox. March 6, pp. 76–81.

Business Week. 1995b. A Bigger—and Cheaper—Online Universe? Sept. 4, p. 100E.

Business Week. 1995c. PCs: The Battle for the Home Front. Sept. 25, pp. 110–114.

Business Week. 1995. Bullet-Proofing the Net. Nov. 13, pp. 98–99.

Business Week. 1996. Wireless' Wild, Wild North. March 11, pp. 87–90.

DOT. 1995a. *Status of the Nation's Surface Transportation System: Condition and Performance.* Report to Congress, Oct.

DOT. 1995b. *1990 NPTS Report Series: Special Reports on Trip and Vehicle Attributes.* Office of Highway Information Management, Federal Highway Administration.

Gates, B. 1995. *The Road Ahead.* Viking, New York.

Hodge, D.C., R.L. Morrill, and K. Stanilov. 1995. *Metropolitan Form Implications of Intelligent Transportation Systems.* Office of Technology Assessment, Congress of the United States; University of Washington, Seattle.

Hoffman, D.L., and T.P. Novak. 1995. *Marketing in Hypermedia Computer-Mediated Environments: Conceptual Foundations.* Working Paper 1, Project 2000: Research Program on Marketing in Computer-Mediated Environments, Vanderbilt University, Nashville, Tenn.

Kelly, K. 1994. *Out of Control.* Addison-Wesley, Reading, Mass.

KSA. n.d. *Vision for the New Millennium.* Atlanta, Ga.

Midwinter, J.E. 1995. Convergence of Telecommunications, Cable, and Computers in the 21st Century: A Personal View of Technology. In *Annual Review of the Institute for Information Studies*, Northern Telecom, Inc., Nashville, Tenn., and The Aspen Institute, Queenstown, Md.

Mokhtarian, P. 1990. A Typology of Relationships between Telecommunications and Transportation. *Transportation Research,* Vol. 24, No. 3, pp. 231–242.

Negroponte, N. 1995. *being digital.* Knopf, New York.

Newsweek. 1995. Special Report on TechnoMania. Feb. 27.

Niles, J.S. 1994. *Beyond Telecommuting: A New Paradigm for the Effect of Telecommunications on Travel.* Office of Energy Research, U.S. Department of Energy.

Pisarski, A.E. 1992. *Travel Behavior Issues in the 90's.* Office of Highway Information Management, Federal Highway Administration, U.S. Department of Transportation.

Sandro, B.D. 1995. Battelle's Best Guesses. *Omni Magazine,* Winter, pp. 42–117.

Time. 1995. Special Issue, Spring.

U.S. Census Bureau. 1995. *Current Population Survey and Computer Ownership/Usage Supplement.* Nov.

Washington Post. 1995a. Billion-Bit Memory Chip Developed by NEC. Feb. 14, p. B2.

Washington Post. 1995b. Inquiring Minds Want To Know the Secrets of Your On-line Life. Dec. 4, Business Section, p. 20.

Washington Post. 1996. Media Giants' Bedfellowship Raises Questions About Competition. Jan. 7, pp. H1–H9.

Appendix D

Ecological, Environmental, and Energy-Related Issues

David L. Greene
Oak Ridge National Laboratory

Transportation is a large and important component of modern economies. The U. S. transportation system generated more than one-tenth of the nation's gross domestic product in 1994, producing 4.4 trillion passenger-mi of travel and 3.5 trillion ton-mi of freight (BTS 1995). In so doing, it consumed an enormous quantity of energy, 23.4 quadrillion BTUs. Highways produce 90 percent of passenger miles, about one-fourth of intercity ton miles, and they account for three-fourths of transportation's energy use (Davis and McFarlin 1996). Such a prodigious amount of activity inevitably creates profound and diverse environmental effects. These effects range from air pollution from the combustion of fuel, to noise pollution created by engines and moving vehicles, to the modification or destruction of habitats by the expansion of transportation infrastructure and its transformational impacts on the use of land. A great many, though by no means all, of transportation's effects on the environment are related to its use of energy in the form of fossil fuels. "Using energy in today's ways leads to more environmental damage than any other peaceful human activity (except perhaps reproduction)" (*Economist* 1991). Thus, transportation's energy and environmental problems are closely related.

Increasingly, the energy and environmental impacts of transportation must be understood on a global scale. Not only in the United States but around the world, motorized transport, with its appetite for fossil fuels and their ensuing environmental impacts, predominates and continues to grow rapidly. For example, in 1950 the United States' 50 million cars and trucks composed

70 percent of the world's motor vehicle fleet (Figure D-1). Today, the United States' 200 million vehicles account for less than one-third of a world vehicle stock that has been growing at an annual rate of 3.3 percent within the United States but at 7.4 percent in the rest of the world (Davis and McFarlin 1996). Transportation energy use has grown fastest in the developing economies of the world, averaging 4.5 percent over the past two decades. Even in Europe and Japan, areas we sometimes look to as models of what the United States might become if land use and transportation systems could be transformed for greater efficiency, motorized transport and its energy use have been growing at faster rates (1.8 percent/year) than in the United States (1.3 percent/year) (Davis and McFarlin 1996). Globally expanding motorized transport has made urban air pollution a worldwide problem, produces about a fifth of the world's greenhouse gas (GHG) emissions, promotes the transformation of landscapes from Indonesia to Brazil, and puts increasing pressure on geographically concentrated sources of the world's oil supplies.

KEY CONCEPTS

Five concepts are key to understanding transportation's ecological, environmental, and energy problems:

1. The pervasiveness of transportation and its effects on human use of the land,
2. The external costs associated with transportation activity and infrastructure,
3. The provision of much transportation infrastructure as public goods,
4. The consequences of energy market defects for transportation, and
5. Uncertainty about the costs to society of environmental damages and petroleum dependence.

Pervasiveness

The first concept is simple but enormously important: wherever people go, transportation must also. By the same token, where transportation infrastructure is created, people and development can and usually will follow. These simple observations have profound implications for the scope and extent of transportation's environmental impacts. Not only can transportation infrastructure itself fragment habitats, occupy wetlands, disturb habitats and

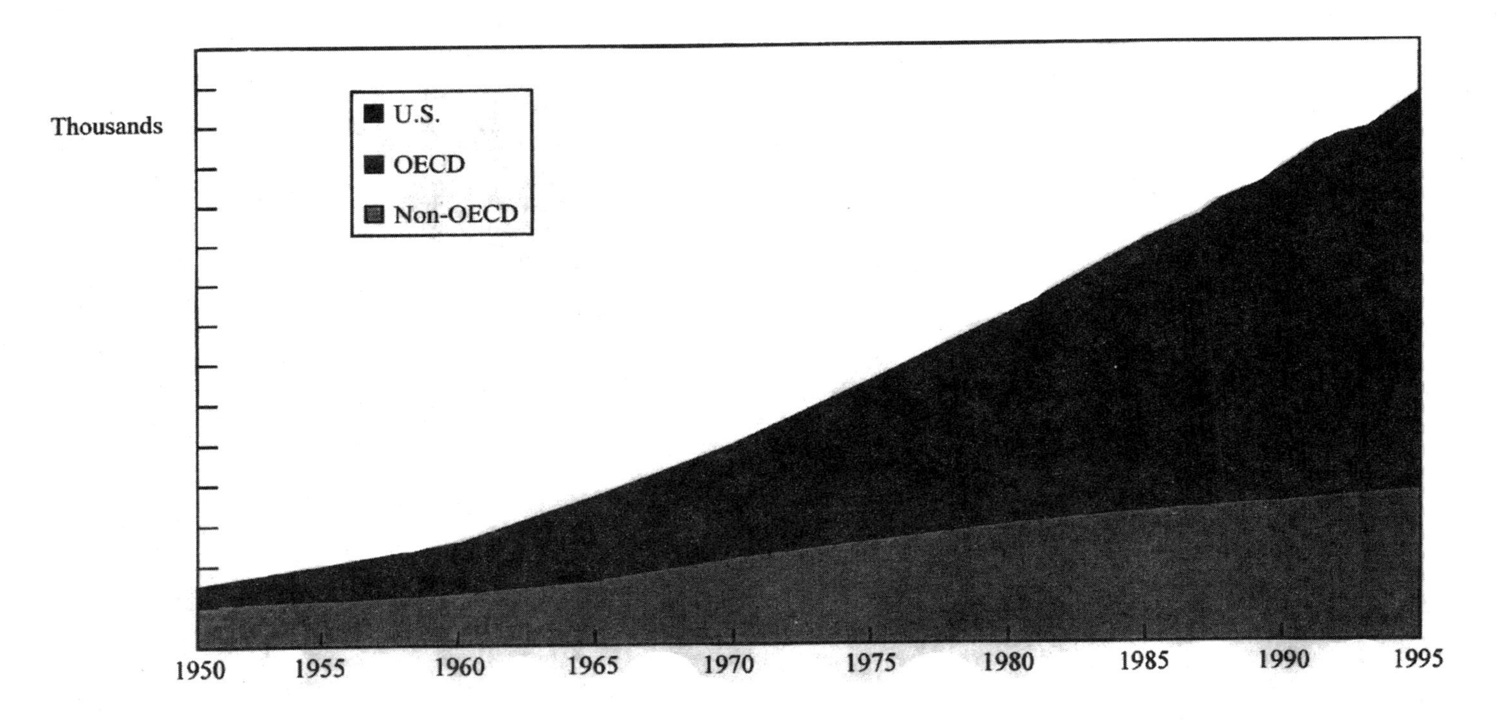

Figure D-1. Motor vehicle stocks: rapid expansion outside the United States.

other sensitive areas, and otherwise directly change the landscape, but the accessibility conferred by transportation infrastructure generally leads to changes in existing land uses. For example, construction of roads has been found to be an essential prerequisite for converting rainforests to commercial agriculture in Brazil (Dale et al.1993). The role of highways in creating more dispersed urban settlement patterns (urban sprawl) is well-known.

External Costs

More often than not, the environmental consequences of transportation are not taken into account in free-market decision making. Because the environment, in essence, belongs to everyone and to no one, individuals acting only in their self-interest will not consider adequately environmental damages in making private decisions. Environmental damages are external to private decision making in free-market economies. As a result, public policy action generally is required to solve any of transportation's environmental problems. A variety of policy strategies are available from regulations to quasimarket mechanisms to research and development (R&D). Which is best will depend on the circumstances. Because of the pervasiveness of transportation and because of its growth around the world, public policy actions to correct transportation's environmental problems will be a continuing, large-scale problem.

Public Goods

The fact that much of transportation infrastructure is provided by governments as public goods amplifies the importance of public policy in addressing the effects of transportation on the environment. The most obvious implication is that public agencies are directly responsible for assessing and mitigating environmental impacts from the expansion and operation of public infrastructure. Equally important, public policy must also consider the transformational effects of infrastructure investments on the landscape. Provision of highway infrastructure as a public good, in particular, has often given rise to recondite arguments about implicit subsidies and hidden costs that are believed to distort transportation behavior.

Energy Market Defects

Motorized transportation relies on petroleum for energy. In the United States, transportation is 97 percent petroleum-dependent. By accident of geology, most of the world's economically exploitable petroleum reserves lie within the borders of a small number of countries that have joined together to form the Organization of Petroleum Exporting Countries (OPEC). Since 1973, when the Arab member states organized a deliberate boycott of the United States and other allies of Israel, oil market disruptions and price shocks have cost oil-consuming economies trillions of dollars (Greene and Leiby 1993). Although oil prices have been low and stable since 1986 and oil supplies have been abundant, little of fundamental importance to the dynamics of world oil markets has changed since the early 1970s. OPEC's loss of market share while defending high prices from 1979 to 1985 is an important exception. But much of its lost market share has already been regained. With the continuing growth of transportation and, thus, world oil demand, the reemergence of oil dependence as a significant economic problem in the coming decade appears likely.

Uncertainty About Social Costs

A combination of scientific uncertainty about the size and seriousness of environmental impacts, together with the difficulty of establishing the value of largely extramarket goods, makes it inherently difficult for society to decide how much of its resources to devote to solving transportation's energy and environmental problems. Global warming is perhaps the best example. Not only is it not clear how much the world's average temperature will rise (current best estimates range from 1.9°F to 6.7°F) (IPCC 1996), but the consequences of such a rise in terms of regional and local weather and climate, sea level changes, and such are understood even less. Social scientists are only beginning to try to evaluate the potential impacts of such changes (IPCC 1996). Even in relatively well-studied areas, such as air pollution by motor vehicles, major uncertainties about impacts remain. For example, re-entrained road dust (small debris kicked up off of roads by the wakes and tires of vehicles) is reported to be the second largest source of small particulate matter in the United States, accounting for more than 40 percent of all PM10 emissions (EPA 1995). Current scientific understanding of the roles of size and chemical composition in determining the health effects of particulates points to road dust as a potentially dangerous air pollutant with significant human mortality and morbidity impacts (Delucchi and McCubbin 1996). Yet there is enough uncertainty that road dust remains an unregulated pollutant. Uncertainty about the value of environmental damages impedes both developing consensus and formulating policies to address transportation's environmental and energy problems.

Taken together, these five factors imply that public policy will be at the heart of solving transportation's energy and environmental problems and that the implementation of effective and efficient public policies will continue to be challenging. Furthermore, as transportation infrastructure and activity continue to grow and become increasingly motorized around the world, the necessity of dealing comprehensively with transportation's environmental and energy problems becomes a global concern.

MAJOR ISSUES

Five major energy and environmental issues are likely to change society's relationship to highway transportation over the next few decades:

1. The continuing struggle to improve urban air quality,
2. The search for appropriate and meaningful policies to mitigate GHG emissions,
3. The likely reemergence of oil dependence as a national economic and security issue,
4. The need to better understand and manage the interrelationships between transportation infrastructure and land use, and
5. The growing movement toward a fundamental integration of ecological, energy, and environmental concerns into the planning and operation of transportation systems to achieve sustainability.

Urban Air Quality

Despite enormous progress in controlling emissions from properly operating vehicles equipped with modern emissions controls, meeting National Ambient Air Quality Standards (NAAQS) remains a nagging concern. What will it take to achieve national goals for air quality?

The case can be made that efforts to control air pollution from transportation vehicles have been enormously successful. On the other hand, it can be said that they have failed. The good news is that, despite a doubling of vehicle miles of travel (VMT) since 1970, transportation emissions of most criteria pollutants have been reduced. Indeed, the U.S. Department of Transportation has estimated that had there been no improvement in the rates of emissions per vehicle mile since 1970, air pollution by transportation would be two to four times what it is today (BTS 1995). A new, properly operating automobile emits less than one-tenth as much pollution as a pre-1967 automobile. Federal emissions standards appear to have been remarkably effective in stimulating advances in pollution control technology and its implementation.

The bad news is that as of February 8, 1996, 182 of the 268 metropolitan statistical areas (MSAs) in which 80 percent of the U.S. population lives failed to attain one or more of the NAAQS (EPA 1994). In nearly all areas there has been some improvement over 1980 air quality levels, yet the result still falls disappointingly short of the federal, state, and local plans that predicted attainment of NAAQS by this time. MSA air quality ratings for ozone are shown in Figures D-2 and D-3. Improvement is evident (Los Angeles, for example, has moved from the "emergency" to the "warning" level), but there are still far too many MSAs whose air is described as unhealthful.

There are two key reasons that so many MSAs still flunk NAAQS despite one to two orders of magnitude reductions in pollutant emissions from new vehicles. First, the demand for vehicle travel grew at 3 to 4 percent per year, and fuel consumption increased as well. In the future, DOT expects highway travel to increase at 2 to 3 percent per year. A second reason is that emissions control technology does not perform as well under real-world conditions as it does on government certification tests. It took some time for the scope of this problem to be measured and understood. It was not until 1991 that the National Research Council observed that real-world emissions levels of key automotive pollutants were probably twice what certification test-based emissions models were predicting (NRC 1991).

Only recently have the causes of the shortfall been identified and measured (Ross et al 1995; German 1995a). "Off-cycle" operation, driving behaviors not contained in the federal test procedure, can temporarily multiply emission rates of carbon dioxide (CO_2) and volatile organic compounds by orders of magnitude. "Super-emitting" vehicles whose emissions control systems do not function effectively due to poor construction, poor maintenance, age, or tampering also appear to be a bigger problem than previously thought. Air conditioner operation now appears to increase nitrogen oxide (NO_x) emissions much more than anticipated. The Environmental Protection Agency (EPA) has already begun revising its regulatory procedures in an attempt to correct these deficiencies (German 1995b).

The 1990 Clean Air Act Amendments (CAAA) introduced a variety of measures to further reduce transportation emissions in pursuit of the NAAQS. The CAAA requires that nonattainment areas implement enhanced

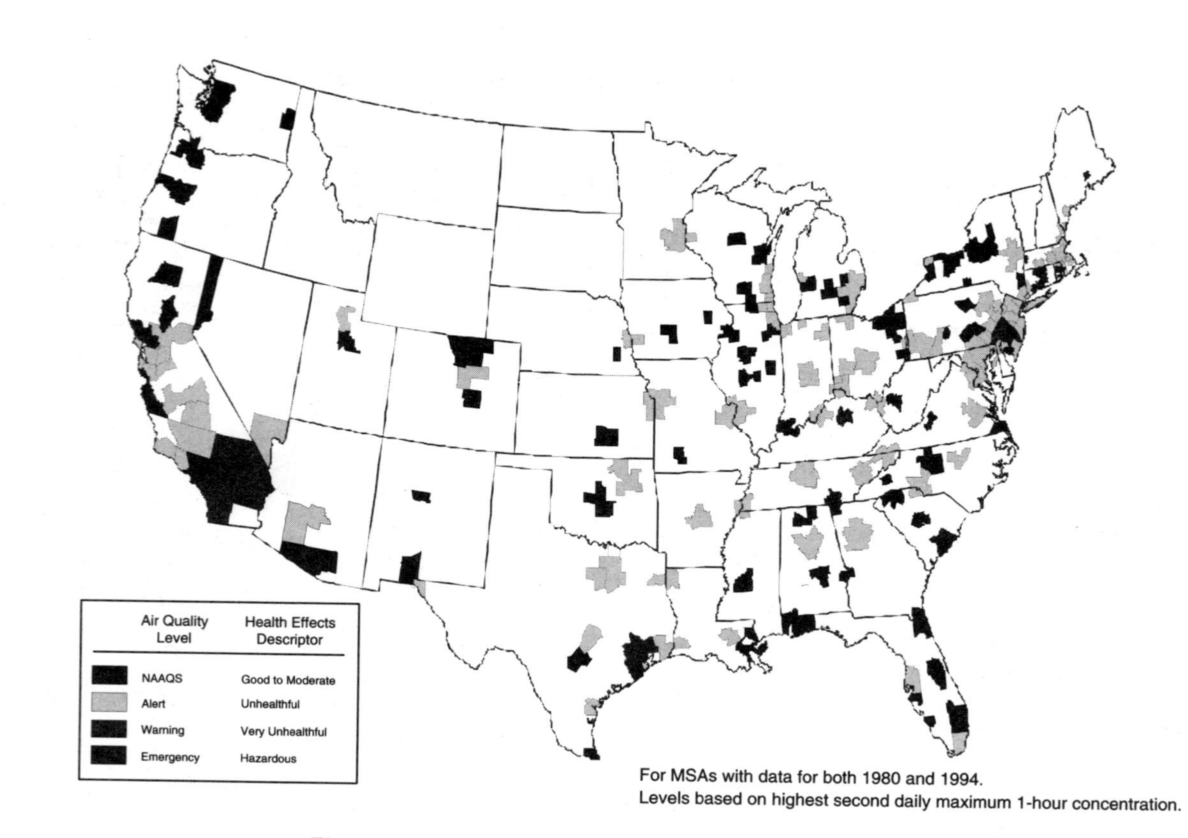

Figure D-2. MSA air quality levels for ozone in 1980.

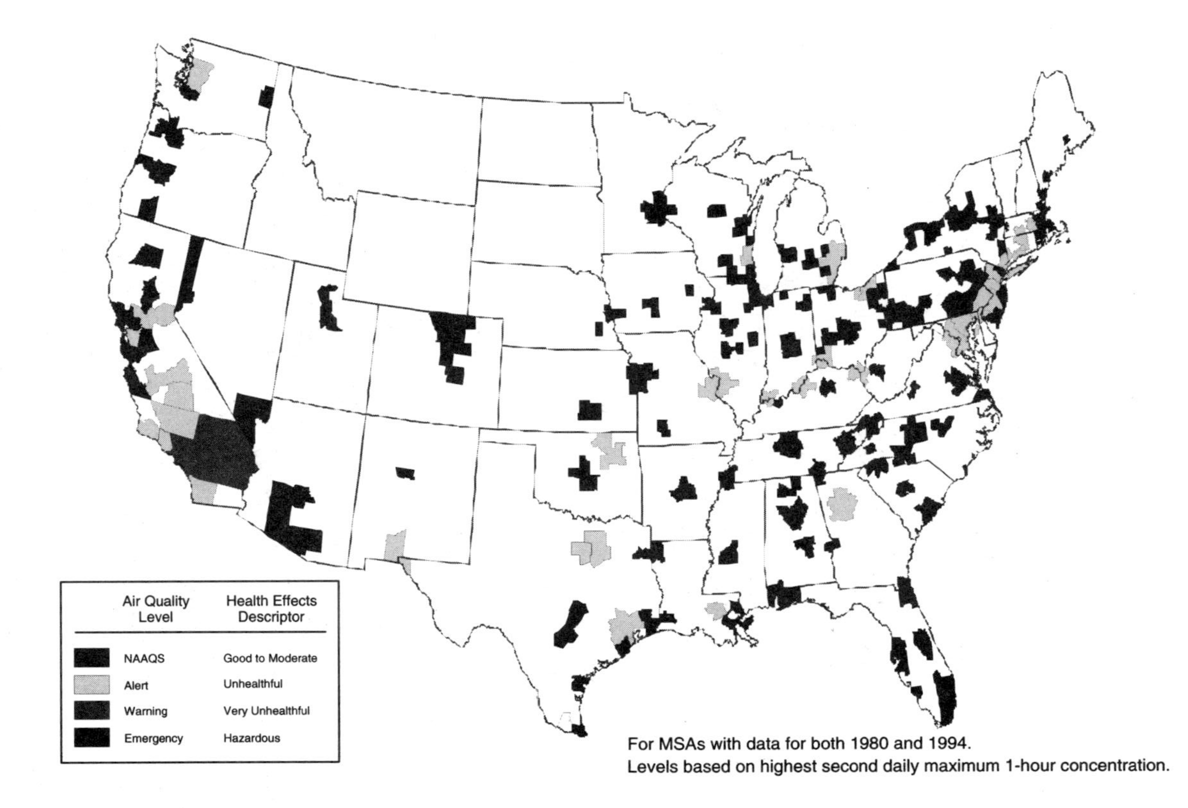

Figure D-3. MSA air quality levels for ozone in 1994.

inspection and maintenance (I/M) programs to identify and correct super-emitting vehicles. EPA has developed a new 240-second test procedure (IM240) to be used as part of enhanced I/M programs; it identifies high emitters far more accurately than previous procedures did. Reformulated gasoline, also required in ozone nonattainment areas, will produce fewer evaporative emissions because of its lower vapor pressure. In addition, it contains oxygenates (alcohols and ethers) to reduce CO emissions, and far lower levels of toxic constituents, such as benzene. Reductions in the sulfur content of gasoline and diesel fuel will cut emissions of sulfate particulates from diesel exhaust and produce modest reductions in all criteria pollutants from catalyst-equipped gasoline vehicles (Walsh 1994). The 1990 CAAA also extends pollution standards to previously unregulated emissions from non-highway modes. Under the 1990 CAAA, California is given authority to implement a comprehensive program combining clean fuels and clean vehicles to achieve greater emissions reductions. California's programs are likely to add preheated catalytic converters, sophisticated onboard diagnostic systems, and even battery electric vehicles to the list of pollution control technologies. Both the 1990 CAAA and the 1992 Energy Policy Act (EPACT) encourage alternative (nonpetroleum) fuel vehicles through a combination of purchase mandates for government and certain other fleets, and incentives for investments in vehicles and refueling infrastructure.

Will these strategies close the gap between test and real-world emissions performance? Will they, even taken together, be enough to win the race against the growth of vehicle travel? If not, still greater technological changes will have to be made to alternative vehicles and fuels and possibly the management of VMT. Will alternative fuels and vehicles be an important part of the solution, or will cleaner conventional vehicles and fuels continue to predominate? A recent report of the Transportation Research Board has questioned the effectiveness and appropriateness of emission control strategies that restrict highway capacity in order to reduce emissions by constraining VMT (TRB 1995). The report suggested that more comprehensive congestion and vehicle travel pricing strategies be considered as an alternative complement to technologically based emissions controls.

GHG Emissions and Global Climate Change

Were it not for the greenhouse effect, the earth's average temperature would be about 59°F colder than it now is, too cold for life as we know it (NRC 1991). There is no scientific debate about the existence of the greenhouse effect or its importance to the biosphere. There is also no doubt that human activities have added substantial amounts of GHGs, especially carbon dioxide (CO_2), to the atmosphere since the onset of the industrial revolution. Annual additions of CO_2 have increased from about 100 million metric tons (MMT) in 1860 to 6200 MMT in 1991 (Keeling 1994). Scientists also know that about half of the tons added have remained in the atmosphere, and that atmospheric CO_2 concentrations have increased from about 280 parts per million by volume (ppmv) in 1750 to 353 ppmv in 1990 (Neftel et al. 1994). Scientists are also quite certain that global temperatures have increased between 0.5°F to 1.1°F during the past 100 years (Jones et al. 1994). Here the scientific certainty ends, however.

It is not clear how much and when temperatures will rise as a result of the continuing buildup of CO_2 from human sources. Scientists do not know the exact relationship between GHG concentrations and average earth temperature. It is not even clear that anthropogenic emissions caused the observed temperature increase; the historical record shows much larger natural temperature fluctuations. Our best models predict that average temperatures will rise by 1.9°F to 6.7°F when atmospheric concentrations reach about 560 ppmv by the end of the next century. But these predictions are enormously uncertain. Still less is known about the details. What weather and climate changes will occur where? What will be the consequences for human activities? The uncertainty makes it that much more difficult to build a global consensus about what, if anything, to do.

One more thing appears certain: solving the climate change problem with today's technology almost certainly costs more than the economies of the world are willing to pay. EPA has estimated that reductions of 50 to 80 percent from current and projected future levels of GHG emissions would be necessary to stabilize atmospheric concentrations at their current levels (NRC 1990). The Intergovernmental Panel on Climate Change estimated that emissions would have to be reduced immediately to 30 percent of current levels to keep atmospheric concentrations constant (IPCC 1990). Such reductions would most likely cost on the order of hundreds of dollars per ton of CO_2 reduced (Nordhaus 1991; Kram and Hill 1996). Because a gallon of gasoline produces about 20 lb of CO_2, 100 gal is roughly equivalent to 1 ton of CO_2. This would be equivalent to paying dollars per gallon of gasoline consumption reduced, over and above the value of the gasoline saved. Put another way, the U.S. economy generated more than 5 billion metric tons of CO_2 in 1994 (DOE 1995). At $100/ton, the cost of solving the U.S. economy's share of the climate change problem right now would be about half a trillion dollars per year, about $2,000/capita. If we are unwilling to pay the cost now, our only serious alternative strategy would appear to be to buy time and try to improve the terms by creating new technological options that reduce the costs of a solution.

A cost-effective solution to this problem of uncertain magnitude apparently does not yet exist, but can we afford to ignore the potential threat? Climate change is truly a global problem that nations are unlikely to solve unilaterally. How will the need to cooperate with the rest of the world in solving this problem affect the U.S. highway system, the greatest producer of GHGs among the world's transport systems?

Oil Dependence

In October 1973, the Arab members of OPEC announced an oil export boycott against countries aiding Israel during the October War. Though the reduction in oil supply from these countries was only about 10 percent or less of total world oil production, world oil prices immediately doubled. Again in 1979, the Iran-Iraq War caused a loss of more than 5 million barrels per day (mmbd) of oil production, about 10 percent of world supply. Oil prices doubled again. In both cases, OPEC members restrained their oil output in succeeding years to sustain the high price levels. After 6 years of cutting back production to maintain price levels and a consequent loss of about two-thirds of their 1973 market share, core members of the OPEC cartel, and especially Saudi Arabia, abandoned their policy of defending higher prices. Prices crashed and have remained low ever since, except in 1990 when loss of supply of nearly 5 mmbd from Iraq and Kuwait during the Persian Gulf War caused prices to jump from $17.50 to $33 per barrel (Greene 1996). Unlike the previous events, the 1990 price shock was short-lived as Saudi Arabia and the United Arab Emirates used their enormous reserve capacities to restore nearly all of the supply shortfall. In 1996 a minor shortage of gasoline stocks caused a gasoline price blip that startled U.S. politicians, at least.

Is the oil problem over? Is OPEC dead? Certainly oil supplies have been abundant and prices low (with two small exceptions) for a decade now. Yet a close look shows that little of fundamental importance to world oil markets has changed. Assessments of proven oil reserves (1 trillion barrels) and ultimately recoverable resources (1.6 trillion barrels) have changed little in the past quarter century (Greene et al. 1995). History suggests that even "ultimate resources" will prove to underestimate what is eventually found and produced. With world oil consumption running at about 25 billion barrels per year, there is oil for at least half a century and almost certainly much more.

Yet oil resources remain as concentrated as ever in the hands of OPEC members, who hold two-thirds to three-fourths of proven reserves and more than half of ultimate resources. Moreover, OPEC is drawing down its reserves at half the rate of the rest of the world. Thus, as world motorization and its appetite for oil grow, more and more of the world's oil will once again be supplied by OPEC. The U.S. Department of Energy projects that between 2000 and 2005 OPEC will regain the more than 50 percent market share it held in the early 1970s. It is very likely that this will make oil dependence once again a serious economic, political, and strategic problem for the United States and other consuming nations.

Outside OPEC's market share, little else important to oil markets has changed. There is no evidence that world oil supply and demand can respond any more rapidly and effectively to a price increase than they did in the 1970s and 1980s. Transportation remains more than 95 percent dependent on petroleum for energy. U.S. oil imports are already within 1 percentage point of their historic high of 46 percent in 1978. Oil cost, as a share of gross national product, is about the same as it was in 1972. The key oil-producing regions of the world remain politically volatile. A careful review of the fundamentals reveals that the oil dependence problem is not dead but only sleeping.

But can this problem be solved by economic or technological policies? The answer is not clear. Several recent studies have demonstrated that the United States' and other countries' strategic petroleum reserves, though useful in brief supply disruptions, are virtually powerless against a determined supply curtailment such as that which occurred in 1973-1974 and 1979-1980 (Suranovic 1994; Greene et al. 1995). What appears to be required is a fundamental change in demand and supply for oil. This means a basic change in the ability to improve transportation energy efficiency and to substitute alternative fuels for petroleum. If such change could be made, it could easily be worth a trillion dollars or more to the U.S. economy over the next 20 years (Greene et al. 1995).

Transportation and Land Use

Transportation affects land use directly when land is used for transportation facilities and indirectly by stimulating development and influencing its intensity and location. Surprisingly little effort has been put into inventorying transportation facilities and their land use, but one source puts U.S. land occupied by road pavement at 20,627 mi^2 (not including unpaved right of way), with parking spaces taking up an additional 2,000 to 3,000 mi^2 (Delucchi 1995). Low-cost, high-speed transportation favors the substitution of land for other factors in production and consumption, thereby magnifying transportation's environmental impacts. Between 1982 and 1992, built-up and urban land uses in the United States increased by 14 million acres, increasing total developed land area to 92.4 million acres, 5 percent of the total land area of the lower 48 states.

Roads and other transportation infrastructure damage natural habitats by destroying habitats, by replacing them with the physical infrastructure itself, disturbing adjacent habitats (e.g., noise, emissions), fragmenting habitats by acting as a barrier to the movement and migration of species, and directly killing animals (e.g., in collisions with vehicles) (Canters and Cuperus 1995). A growing body of research is documenting the effects of transportation systems on natural habitats, particularly in Europe, where habitat loss and fragmentation due to transportation infrastructure is already recognized as a significant problem (Seiler and Eriksson 1995). Ecological engineers are developing methods for mitigating these impacts, such as by making roads and other barrier-like infrastructure permeable to species, and planning procedures are beginning to require restitution or "in-kind" compensation for the loss of wetlands and other critical habitats to transportation infrastructure. As infrastructure for motorized transport continues to expand worldwide, concern over the direct and indirect impacts of transportation on habitats and landscapes will certainly increase.

Sustainability, Full Social Costs, and Integrated Planning and Management of Public Systems

Three incipient "movements" in transportation policy and planning may signal the advent of a paradigm shift in the way that public agencies make transportation investments, and manage and regulate both public and private systems. These three movements, which appear to share a common concern about energy and the environment, are (*a*) sustainability of transportation (OECD 1995), (*b*) full social cost pricing of transportation (DeCorla-Souza et al. 1995), and (*c*) integrated transportation and land use planning (Nelson and Shakow 1994). Each approaches the problem of protecting the environment in the face of ever-increasing demands for mobility from a somewhat different perspective, yet each attempts to forge a comprehensive, integrated strategy.

The increasing complexity, pervasiveness, and interrelatedness of society's energy and environmental problems have prompted some to propose a new goal for human systems: sustainability. The World Commission on Environment and Development has defined sustainable development as "development that meets the needs of the present without compromising the ability of future generations to meet their own needs." While this is far from an operational definition that could be translated directly into transportation policy, it clearly expresses an intent to ensure that the ability of the earth to support human life not be compromised and a concern that human systems now pose a threat to that capability.

All three movements, should they gain momentum, will create a growing impetus to manage highways in a socially optimal manner, or at least accounting for as many external costs as is practical. Advanced information and telecommunications technologies will increasingly make such things feasible. The question will be whether the public will accept such policies.

SOURCES OF AND POTENTIAL FOR MAJOR CHANGES

If the past is a guide to the future, technology will be the major impetus for changing transportation's relationship to the environment. Technology has played the key role in mitigating conventional air pollutant emissions and is likely to continue to do so in the future. Technology has played a relatively minor role to date in mitigating transportation's impacts on land use and habitats. Market mechanisms—attempts to price or create markets for transportation's external costs—have been all but untried in the United States. Developing innovative pricing mechanisms may be essential if transportation's externalities (including traffic congestion) are to be fully addressed. Intelligent transportation systems (ITS) technology will make such real-time user charges feasible, if the public will accept extensive pricing of highway mobility.

In the near term, technological solutions to air pollutant emissions problems abound. A great deal can be accomplished by closing "loopholes" in EPA's test-cycle procedures. Revising emissions tests to include the high speeds and hard acceleration episodes often seen in real-world driving will force manufacturers to abandon the strategy of "command enrichment," a fuel metering strategy that currently increases emissions by 30 percent to more than 200 percent (Ross et al. 1995). This can be accomplished by simply reprogramming vehicles' computer control systems. CAAA-mandated inspection and maintenance and on-board diagnostic requirements, together with extended manufacturer responsibility for the performance of emissions control systems, will also help to improve the real-world performance of previously implemented technological solutions to vehicle emissions. Reformulated gasoline and preheated catalysts will help further reduce emissions from gasoline vehicles. Low-sulfur diesel and lean-NO_x catalysts are likely to help clean up the diesel engine's act.

But continuing growth in vehicle travel of 2 to 3 percent per year will make air quality goals a moving target. A still cleaner generation of vehicles will be required to accommodate yet another doubling of vehicle travel. It is not yet clear that conventional, petroleum-based internal combustion vehicles cannot meet that challenge, but longer-term solutions to motor vehicle emissions may require novel technologies such as alternative fuels, hybrid vehicles, battery electric vehicles, or even fuel cell vehicles. The large-scale alternative-fuel mandates of the EPACT, together with California's clean fuels and vehicles program under the 1990 CAAA, are already beginning to give alternative-fuel vehicles (AFVs) a presence on American highways. From 1992 to 1995, the number of AFVs in the United States increased from 251,470 to 362,190, and the Energy Information Administration estimates that 1996 will see 421,294 AFVs in use (DOE 1996a). The types of AFVs are becoming more diverse, as well. In 1992, 221,000 of the 251,000 AFVs were liquefied petroleum gas (LPG) vehicles. Natural gas, alcohol fuel, and electric vehicles combined amounted to only 30,000. By 1996, more than 140,000 non-LPG AFVs will be in use, an annual growth rate of 47 percent.

A great deal can also be done with conventional technology to reduce GHG emissions from highway vehicles. On a full fuel cycle basis, near-term alternative fuels have been shown to offer little advantage over gasoline and diesel fuels for GHG reduction (Delucchi 1991; DOE 1996b). Thus, in the near term, GHG reduction must be achieved by efficiency improvements. A recent study by the now-defunct Congressional Office of Technology Assessment concluded that new passenger cars could average 50 mpg by 2015 using advanced, conventional drive trains. Hybrid vehicle designs, including battery–fuel cell hybrids, such as are the goal of the government/industry Partnership for a New Generation ofVehicles (PNGV 1994), could push that limit over 80 mpg (OTA 1995). But conventional technology does not appear to offer the kinds of CO_2 reductions that appear to be needed to solve the global climate change problem.

Ultimately, controlling global climate change is likely to require a technological revolution: not just in vehicle technology but in power generation, as well. Furthermore, if the global climate change problem must be solved by radically reducing CO_2 emissions, this revolution is likely to have to be negotiated with the other nations of the world, a novel proposition for the U.S. transport system. New vehicle technology, new infrastructure for new energy sources, and perhaps even new ways of managing highways may be required. Vehicles achieving over 80 mpg are likely to weigh less than 2,000 lb, as compared with a current average weight of more than 3,100 lb. The safety of lightweight, super-efficient vehicles could become a major issue for highway design. Questions of the mix of vehicles in traffic and the aggressiveness of vehicle designs will become more important. New rules and methods of managing traffic streams may be needed.

Much of the current public policy addressing transportation's energy and environmental problems relies on regulations intended to compel technological solutions. As energy and environmental problems become more pervasive and more global, policy makers are likely to seek solutions that are more efficient, comprehensive, and integrated. This will make technological fixes less likely to be employed as a sole strategy. It will make it more likely that market mechanisms will become integral to public policy solutions. Market mechanisms may become attractive for addressing problems of transportation externalities, which include congestion as well as environmental pollution. The incipient movements for full social cost pricing, sustainable transport, and integrated transportation and land use planning are all likely to drive toward greater use of pricing mechanisms to manage transportation externalities.

Pricing transportation is an area in which technological advances and the need to solve nagging congestion and environmental problems may be converging. An issue receiving more attention in Europe than in the United States is whether transportation's congestion, energy, and environmental problems can be solved neatly once and for all by internalizing all externalities (OECD 1995; Button and Nijkamp 1994; Kågeson 1993). While it is doubtful that this holy grail of economic efficiency can ever be achieved completely, advanced information technologies such as envisioned in the ITS will make it more efficient to monitor and charge users in complex ways (TRB 1994). Practical limitations suggest this will be more effective in some areas than in others. For instance, it may never be practical to measure the emissions from each vehicle in real time and assess an optimal pollution charge, taking into account ambient conditions such as weather, circulation, and existing concentrations of airborne pollutants. It should certainly be possible to assess approximately correct congestion charges and average per-mile pollution damage costs.

Whether American motorists will accept per-mile charges that are high enough to restrict their mobility is an open question. Wachs (1994) argues that because there are not powerful interest groups to promote ideas such as congestion pricing and there are many potential opponents in the driving public, congestion pricing schemes will probably never be implemented. Indeed, unless the revenues from congestion pricing schemes are redistributed to the public, all but the wealthiest drivers are likely to feel worse off after than before efficient congestion pricing (Anderson and Mohring 1995). Congestion pricing raises difficult equity issues, as well, because those whose time has the least monetary value are likely to be hurt the most unless they are compensated by redistributed revenues. Thus, achieving political consensus rather than technical and economic practicality will be the sticking point for externality pricing.

IMPLICATIONS FOR RESEARCH AND DEVELOPMENT BY THE FEDERAL HIGHWAY ADMINISTRATION

The implications for the Federal Highway Administration's (FHWA's) R&D efforts of energy and environmental issues are many. There are certain to be significant changes in the technology of highway vehicles, in the way highway systems are operated, and in the way indirect impacts and externalities of highway travel are accounted for and managed.

The use of ITS technology for pricing congestion and other external costs has great potential but considerable barriers. Understanding the conditions under which externality pricing policies can be acceptable to the public is of central importance. Better understanding of the "full costs" and equity issues is also important to formulating policies for road pricing and building a consensus about the role of externality pricing in American highways.

Currently, the Departments of Commerce and Energy have the lead roles in developing nonpolluting advanced vehicle technologies and alternative energy sources. FHWA may or may not wish to become more involved in R&D programs such as the PNGV. However, these programs raise important issues with respect to safety and highway design for new vehicles. For example, if the average passenger vehicle weighs less than 2,000 lb, is it safe for it to occupy the same roadways with heavy-duty vehicles? Will barriers and fixed objects need to be made friendlier? These and other issues pertaining to the compatibility of new technologies and fuels with existing highway infrastructure will certainly demand FHWA's attention.

FHWA has a historical role in modeling real-world emissions and controlling traffic to reduce them. Technological changes in vehicles will surely alter the relationships between traffic flow and vehicle emissions. The existing relationships are just beginning to be understood so that they can be modeled accurately. Proposed rulemakings by EPA to change federal test procedures will greatly change these relationships. Maintaining accurate emissions models will be a challenging research task.

Transportation land use interactions must be understood to enable integrated transportation land use planning, understand the impacts of full social cost pricing, and develop sustainable transportation strategies. Understanding transportation's direct and indirect land use impacts will also require continuing attention. Developing useful, practical integrated land use and transportation forecasting tools remains a challenge.

The U. S. government is likely to be negotiating international agreements to control GHG emissions. These agreements could have profound effects on highway transport. The need to be a player in these ongoing negotiations will also affect FHWA's research needs.

REFERENCES

Abbreviations

BTS	Bureau of Transportation Statistics
DOE	U.S. Department of Energy
EPA	Environmental Protection Agency
IPCC	Intergovernmental Panel on Climate Change
NRC	National Research Council
OECD	Organization for Economic Cooperation and Development
OTA	Office of Technology Assessment
TRB	Transportation Research Board

Anderson, D., and H. Mohring. 1995. Congestion Costs and Congestion Pricing. Presented at the Conference on Measuring the Full Social Costs and Benefits of Transportation, Beckman Center, Irvine, Calif., July.

Button, K.J., and P. Nijkamp. 1994. Transport Externalities, *Transportation Research*, Vol. 28A, No. 4.

BTS. 1995. *National Transportation Statistics 1996*. Report DOT-BTS-VNTSC-95-4. U.S. Department of Transportation.

Canters, K.J., and R. Cuperus. 1995. Fragmentation of Bird and Mammal Habitats by Roads and Traffic in Transport Regions. *International Conference on Habitat Fragmentation, Infrastructure, and the Role of Ecological Engineering*, Ministry of Transport, Public Works and Water Management, MECC, Maastricht, the Netherlands, Sept.

Dale, V.H., F. Southworth, R.V. O'Neill, A. Rosen, and R. Frohn. 1993. Simulating Spatial Patterns of Land Use Change in Rondonia, Brazil. *Lectures on Mathematics in the Life Sciences,* Vol. 23, pp. 29–55.

Davis, S.C., and D. McFarlin. 1996. *Transportation Energy Data Book*, 16th ed. Report ORNL-6898, Oak Ridge National Laboratory, Tenn.

DeCorla-Souza, P., B. Gardner, J. Everett, and M. Culp. 1995. Total Cost Analysis: An Alternative to Benefit Cost Analysis in Evaluating Transportation Alternatives. *Transportation*. Vol. 24, pp. 107–123.

Delucchi, M.A. 1991. *Emissions of Greenhouse Gases from the Use of Transportation Fuels and Electricity*, Vol. 1. Report ANL/ESD/TM-22. Argonne National Laboratory, Ill., Nov.

Delucchi, M.A. 1995. Bundled Private Sector Costs of Motor-Vehicle Use. In *The Annualized Social Costs of Motor-Vehicle Use in the U.S., Based on 1990–91 Data*. Report 6. Institute of Transportation Studies, University of California, Davis.

Delucchi, M.A., and D. McCubbin. 1996. *The Motor Vehicle Fraction of Total Anthropogenic Air Pollution*. Report UCD-ITS-RR-95-15 (16). Institute of Transportation Studies, University of California, Davis, Feb.

DOE. 1995. *Emissions of Greenhouse Gases in the United States, 1987–1994*. Report DOE/EIA-0573(87-94). Energy Information Administration.

DOE. 1996a. *Alternatives to Traditional Transportation Fuels 1994*. Report DOE/EIA-0585(94)/1. Energy Information Administration.

DOE. 1996b. *Market Potential and Impacts of Alternative Fuel Use in Light-Duty Vehicles: A 2000/2010 Analysis, Technical Report 14: Assessment of Costs and Benefits of Flexible and Alternative Fuel Use in the U.S. Transportation Sector.* Report DOE/PO-0042. Office of Policy and Office of Energy Efficiency and Renewable Energy.

Economist. 1991. August 31, p. 3.

EPA. 1994. *National Air Quality and Emissions Trends Report, 1993*. Report EPA-454/R-94-031. Office of Air Quality Planning and Standards, Research Triangle Park, N.C. [www.epa.gov/airs/nonattn.html].

EPA. 1995. *National Air Pollutant Emissions Trends, 1900–1994.* Report EPA-4564/R-95-011. Office of Air Quality Planning and Standards, Research Triangle Park, N.C., Oct.

German, J. 1995a. Off-Cycle Emissions and Fuel Efficiency Considerations. Presented at 1995 Conference on Sustainable Transportation Energy Strategies, Asilomar Conference Center, Pacific Grove, Calif.

German, J. 1995b. *Observations Concerning Current Motor Vehicle Emissions.* Technical Paper Series 950812, Society of Automotive Engineers, Warrendale, Pa.

Greene, D.L. 1996. *Transportation and Energy.* Eno Foundation for Transportation, Lansdowne, Va.

Greene, D.L., and P.N. Leiby. 1993. *The Social Costs to the U.S. of Monopolization of the World Oil Market.* Report ORNL-6744, Oak Ridge National Laboratory, Tenn.

Greene, D.L., D.W. Jones, and P.N. Leiby. 1995. *The Outlook for U.S. Oil Dependence.* Oak Ridge National Laboratory, Tenn.

IPCC. 1990. *Climate Change: The IPCC Scientific Assessment.* Cambridge University Press.

IPCC. 1996. *Climate Change 1995: Impacts, Adaptations and Mitigation of Climate Change: Scientific-Technical Analysis.* Working Group II. Cambridge University Press.

Jones, P.G., T.M.L. Wigley, and K.R. Briffa. 1994. Global and Hemispheric Temperature Anomalies—Land and Marine Instrument Records. In *Trends '93: A Compendium of Data on Global Change* (T.A. Boden, D.P. Kaiser, R.J. Sepanski, and F.W. Stoss, eds.), Report ORNL/CDIAC-65. Carbon Dioxide Information Analysis Center, Oak Ridge National Laboratory, Tenn., pp. 603-608.

Kågeson, P. 1993. *Getting the Prices Right: A European Scheme for Making Transport Pay Its True* Costs. European Federation for Transport and Environment, Stockholm, Sweden.

Keeling, C.D. 1994. Global Historical CO_2 Emissions. In *Trends '93: A Compendium of Data on Global Change* (T.A. Boden, D.P. Kaiser, R.J. Sepanski, and F.W. Stoss, eds.), Report ORNL/CDIAC-65. Carbon Dioxide Information Analysis Center, Oak Ridge National Laboratory, Tenn., pp. 501–504.

Keeling, C.D., and T.P. Whorf. 1994. Atmospheric CO_2 Records from Sites in the SIO Air Sampling Network. In *Trends '93: A Compendium of Data on Global Change* (T.A. Boden, D.P. Kaiser, R.J. Sepanski, and F.W. Stoss, eds.), Report ORNL/CDIAC-65. Carbon Dioxide Information Analysis Center, Oak Ridge National Laboratory, Tenn., pp. 16–26.

Kram, T., and D. Hill. 1996. A Multinational Model for CO_2 Reduction. *Energy Policy*, Vol. 24, No. 1, pp. 39–51.

Marland, G., R.J. Anders, and T.A. Boden. 1994. Global, Regional and National CO_2 Emissions. In *Trends '93: A Compendium of Data on Global Change* (T.A. Boden, D.P. Kaiser, R.J. Sepanski, and F.W. Stoss, eds.), Report ORNL/CDIAC-65. Carbon Dioxide Information Analysis Center, Oak Ridge National Laboratory, Tenn., pp. 505–584.

Neftel, A., H. Friedl, E. Moor, H. Lotscher, H. Oeschger, V. Siegenthaler, and B. Stauffer. 1994. Historical CO_2 Record from the Siple Station Ice Cove. In *Trends '93: A Compendium of Data on Global Change* (T.A. Boden, D.P. Kaiser, R.J. Sepanski, and F.W. Stoss, eds.), Report ORNL/CDIAC-65. Carbon Dioxide Information Analysis Center, Oak Ridge National Laboratory, Tenn., pp. 11–14.

Nelson, R. and D. Shakow. 1994. *Applying Least Cost Planning to Puget Sound Regional Transportation.* Institute for Transportation and the Environment, Seattle, Wash., Feb.

Nordhaus, W.D. 1991. The Cost of Slowing Climate Change: A Survey. *Energy Journal*, Vol. 12, No. 1, pp. 37–65.

NRC. 1990. *Confronting Climate Change: Strategies for Energy Research and Development.* Energy Engineering Board, Washington, D.C.

NRC. 1991. *Rethinking the Ozone Problem in Urban and Regional Air Pollution.* Committee on Tropospheric Ozone Formation and Measurement, Washington, D.C.

OECD. 1995. *Urban Travel and Sustainable Development.* European Conference of Ministers of Transport, Paris.

OTA. 1995. *Advanced Automotive Technology: Visions of a Super-Efficient Family Car.* Report OTA-ETI-638. U.S. Government Printing Office, Washington, D.C., Sept.

PNGV. 1994. *PNGV Program Plan.* U.S. Department of Commerce, July.

Ross, M., R. Goodwin, R. Watkins, M. Wang, and T. Wenzel. 1995. *Real-World Emissions from Model Year 1993, 2000 and 2010 Passenger Cars.* American Council for an Energy Efficient Economy, Washington, D.C.

Seiler, A., and I.M. Eriksson. 1995. New Approaches to Integrate Landscape Ecological Concepts in Road Planning in Sweden. *International Conference on Habitat Fragmentation, Infrastructure, and the Role of Ecological Engineering,* Ministry of Transport, Public Works and Water Management, MECC, Maastricht, the Netherlands, Sept.

Suranovic, S.M. 1994. Import Policy Effects on the Optimal Oil Price. *Energy Journal,* Vol. 15, No. 3, pp. 123–144.

TRB. 1994. *Special Report 242: Curbing Gridlock: Peak-Period Fees To Relieve Traffic Congestion.* National Research Council, Washington, D.C.

TRB. 1995. *Special Report 245: Expanding Metropolitan Highways: Implications for Air Quality and Energy Use.* National Research Council, Washington, D.C.

Wachs, M. 1994. Will Congestion Pricing Ever Be Adopted? *Access,* No. 4, pp. 15–19.

Walsh, M.P. 1994. Transport and the Environment: Challenges and Opportunities Around the World. *Science of the Total Environment,* No. 146/147, pp. 1–9.

Appendix E

Forecasting Vehicle and Fuel Technologies to 2020

Daniel Sperling
Institute of Transportation Studies
University of California, Davis

As transit use and ridesharing continue their steady decline, motor vehicles are becoming more dominant than ever. They are also becoming larger, increasingly powerful, and more laden with accessories and conveniences. One adverse consequence of motor vehicle proliferation—air pollution—is being mitigated by a continuing stream of technological enhancements, while concern for other consequences, especially petroleum consumption and greenhouse gas (GHG) emissions, languishes. What will be the response to continuing calls for still cleaner air, and episodic (and perhaps intensified) concern over growing petroleum imports and global climate change? Extraordinary consumer wealth in the United States, combined with a veritable revolution in automotive technology, creates the potential for a large array of responses. As the magnitude and potential effectiveness of these technologies become appreciated more widely, the well-documented hesitancy of U.S. political leaders to reduce the harmful consequences of vehicles by restricting their use is likely to be still further weakened.

In this paper, the author focuses on the role of air quality and energy in the design and commercialization of vehicles and fuels. Other adverse consequences—such as noise, land consumption, and aesthetics—are unlikely to play as central a role in the evolution of vehicles.

Today, virtually all motor vehicles are powered by internal combustion engines (ICEs) and petroleum fuels: larger vehicles generally burn diesel fuel in compression ignition engines, whereas lighter vehicles tend to burn gasoline in spark ignition engines. But to what extent and in what way will energy and environmental concerns alter these patterns? Because more stringent emission standards are in place and good progress is being made in achieving them, one outcome is highly certain: emissions of conventional pollutants will continue to decline. What is less certain is whether regulatory and legislative initiatives will force a reduction in fuel use or a shift away from petroleum fuels and ICEs. This paper focuses on how, when, and where these changes may occur, and the implications of those changes for the transportation sector.

ICE VEHICLES

Even if ICE technology is retained, extensive changes are likely, though the implications of these changes for users, the environment, and society would be modest. For instance, continued modifications of gasoline and diesel fuel composition are likely, as refiners and regulators search for the optimal trade-off between emissions and cost. Refiners already have been modifying fuels for years as a means of reducing lead levels, increasing octane, adapting temporally and geographically to different climates, and responding to the needs of electronic fuel injection. Since 1990 these efforts have been accelerated as a result of regulatory requirements for reduced emissions from gasoline and diesel fuel. Future petroleum fuels will have varying amounts of oxygenated compounds and other components, with major implications for refiner investment, but little effect on vehicle users and suppliers.

Greater changes are likely with the engines and vehicles. For instance, huge investments are being made in electronics for the following purposes: safer operation, lower emissions, greater energy efficiency, route navigation, emergency notification, and enhanced accessories. Likewise, the use of lightweight materials, especially lighter steels and aluminum, continues to grow. More composite materials are also being used, but high costs still limit their use (NRC 1996).

The effect of these many innovations on fuel consumption is difficult to predict. Certainly, the energy efficiency of vehicles will continue to improve. Whether these efficiency gains will be translated into fuel economy gains is uncertain. For instance, from 1986 to 1996 average vehicle weight increased 8 percent (from 3041 to 3285 lb) and average acceleration improved 23 percent (from 13.2 sec for 0 to 60 mph to 10.7 sec), but fuel consumption per mile barely changed (Heavenrich and Hellman 1996). The fuel efficiency gains were offset by growing sales of larger vehicles (more than 44 percent of new light vehicles are vans, pickups, and sport utility trucks), increasing power, and more fuel-consuming features such as four-wheel drive (NRC 1996). Sales of light trucks, which exceed 40 percent of the light-duty market, are forecasted to continue increasing, possibly reaching 50 percent in the early years of the next century. The result is that, despite the veritable revolution in electronics and materials, the overall fuel economy of light-duty vehicles is unlikely to improve much, if at all, in the near future. With the slowly expanding use of vehicles, total fuel consumption will also probably expand.

The use of intelligent transportation system (ITS) technologies is not expected to alter this trend much. Improved traffic management and greater availability of traffic and route information will reduce delay due to incidents, but there has been no evidence to suggest that total vehicle travel, energy use, or emissions would be affected substantially by deploying these technologies. Indeed, greater ease of travel may result in more travel. This would almost definitely be the case with automated controls, and possibly with traffic management and information technologies as well. The only prospect for substantial reductions in travel, energy use, or emissions is the application of ITS technologies to paratransit and ridesharing, but because current trends are in the opposite direction—away from transit—most analysts are skeptical of the potential for increased transit and paratransit use, even with smart paratransit and ridesharing. The institutional barriers arrayed against paratransit and strong security concerns of individuals are acting to undermine efforts to expand this intermediate mode.

If fuel consumption is to be reduced, it will require some mix of technology changes, alterations in consumer preferences, and government action to alter market signals and corporate behavior. In the United States, government has been averse to discouraging fuel and vehicle use by consumers, and since the late 1980s has become more reluctant to impose direct fuel-economy restrictions on vehicle suppliers. With low oil prices, fading public concern for energy security, and apathy toward climate change, consumers have lost interest in fuel economy. That could change, with dramatic effects on fuel consumption. For instance, if consumers were to roll back their expectations for power and interior space to levels witnessed in the early 1980s—when light trucks accounted for only about 30 percent of light-duty vehicle sales and acceleration times for 0 to 60 mph were 40 percent less (Heavenrich and Hellman 1996)—then considerably improved fuel economy and modest overall reductions in light-duty vehicle fuel use would be possible. But as automaker fears of more stringent fuel economy standards recede, and as demand continues to grow for the least-efficient vehicles (light trucks), automaker investments are directed away from those technologies and products that could lower fuel consumption.

The trend toward increased fuel consumption by heavy trucks is even stronger. From 1970 to 1993, fuel economy for medium and heavy-duty trucks improved only 15 percent, from 4.8 to 5.5 mpg (Greene 1996). These small improvements were swamped by huge increases in truck travel, as well as other factors such as greater traffic congestion. As a result, energy use by trucks has grown rapidly and is not expected to slow. For instance, from 1970 to 2010, energy use by light-duty vehicles is expected to increase 44 percent (11 percent from 1990 to 2010), compared with 200 percent for heavy-duty vehicles (trucks over 10,000 lb gross vehicle weight plus buses) (Mintz and Vyas 1993).

With dim prospects for expanded transit use and ridesharing, and with increasing car and truck use, all signs point to heightened dependence on private motor vehicles. As indicated in the following, with trends toward larger and more powerful vehicles, one cannot escape the conclusion that technology strategies for reducing fuel use and GHG will become even more central.

The transition to new fuels and vehicle technologies is less predictable than it might be for many other new technologies. That is because motor vehicles have a large effect, both real and perceived, on a variety of energy and environmental concerns. But because these concerns are mostly outside the marketplace, they are manifested principally through government action. Thus, vehicle design and fuels marketing are subject not only to the normal uncertainties of innovation and market demand, but also to the vagaries of government action. Understanding the relative influence and salience of the various energy and environmental concerns provides considerable insight into the evolution of propulsion technologies and fuels.

UNCERTAIN AIR QUALITY INFLUENCES

Government has been more aggressive and effective in curtailing air pollution than any other energy or environmental concern. This focused commitment is likely to waver, since most metropolitan areas—with the notable exception of most of California and a few other major regions—are expected to attain federal ambient air quality standards within the next decade. Because clean air has such strong public support, and air quality laws and rules have such strong enforcement provisions built into them, public interest groups have used air quality concerns as a surrogate for a raft of other environmental, energy, and social concerns, including "livable cities," urban sprawl, and decay of urban downtowns. But will curtailed federal commitment to clean air slow efforts to create more benign motor vehicles? Probably not, for three reasons.

First, continued growth in population and vehicle usage in many regions of the country will forestall federal efforts to absolve itself of responsibility. As air pollution concentrations continue to drop, local regulators will probably dispense with the less effective strategies—especially those aimed at reducing travel—in favor of more effective technology-based initiatives.

Second, California, where air pollution is a permanent problem, has always been the international leader in reducing vehicle pollution; recent examples of California being imitated include reformulated gasoline and tightened "low-emission vehicle" standards. Even where the federal government has not consciously imitated California, it has adopted initiatives intended to provide a relief valve, as with the Partnership for a New Generation of Vehicles, or indirect support, as with the U.S. Advanced Battery Consortium (US ABC). The US ABC, funded half by the U.S. Department of Energy (DOE) and half by industry, has provided more than $200 million in the past few years to develop advanced batteries for use in the zero emission (battery-powered electric) vehicles (ZEVs) mandated by California. The PNGV, a loosely organized program of the Big Three automakers and the Clinton Administration, was in part a desire to develop an environmentally attractive alternative to battery electric cars (i.e., California's mandated ZEVs) that has performance attributes more comparable to those of conventional cars.

Because of the continuing air pollution problems in California, the state will probably continue to pursue clean air aggressively through zero emission technology. On March 28, 1997, it softened the ZEV mandate, adopted earlier in 1990, by eliminating the requirement that 2 percent of vehicle sales be zero emitting in 1998 and 5 percent in 2001, instead requiring that the seven largest marketers of cars in California implement what is essentially a very large demonstration of advanced electric vehicle (EV) technology. The requirement for 10 percent ZEVs in 2003 was retained. Whether the 10 percent requirement continues to be retained depends in large part on automaker success in building and marketing EVs during the rest of the decade.[1]

Third, air pollution levels are increasing and becoming of greater concern in many cities of Europe and Asia. As a result, other nations and European and Japanese automotive companies are stepping up their investments in fuel cells, hybrid electric drivelines, and other very clean (and efficient) propulsion technology. The unveiling by Mercedes-Benz of a fuel cell car on May 14, 1996, and its announcement that it may be ready to sell fuel cell cars by 2010, indicates global demand for environmentally benign vehicles and the intention of international companies to supply that technology.

In summary, because of its continuing air pollution problems, the large size of its market, and recognized leadership role, California will most likely be effective in continuing to stimulate investments in EV (and other electric-drive) technology. Growing concern for air pollution elsewhere in the world will probably strengthen support for these advanced technologies. However, momentum will likely slow, in the United States at least, if air pollution continues to be the sole policy justification.

LANGUISHING ENERGY AND ENVIRONMENTAL CONCERNS

Public demand for reductions in imported oil, GHG emissions, and other environmental impacts such as noise have been muted. Energy security arouses occasional interest, but government initiatives since the early 1980s have become weaker. Perhaps the only notable action has been to increase corporate average fuel economy (CAFE) standards for light trucks by a small 0.1 mpg per year, but in 1996 Congress called for an end to even these small increases.

Government intervention on behalf of other concerns has been even less visible. Although the Clinton Administration signed an international agreement to reduce GHG emissions to 1990 levels and prepared a "Climate Action Plan," the only substantive GHG reduction initiative aimed at the transport sector launched in recent years has been the PNGV. Meanwhile, a high-level advisory board to the President (known as "Car Talk"), estab-

[1] New York and Massachusetts, the only other states with mandates for ZEV sales requirements, did not drop their 2 percent ZEV sales requirement and have thus far survived court challenges from the automobile manufacturers.

lished to create a consensual plan of action, disbanded in fall 1995 for lack of consensus.

Whether these non-air quality goals will gain strength is unknown. Public concern for these other goals appears to be far off. Indeed, in recent years, vehicle travel and fuel consumption have increased substantially—personal vehicle travel, mostly because of less ridesharing, greater participation by women in the market economy, and increasing suburbanization; and fuel consumption, because of more travel, more trucks, flat fuel economy, and more air travel.

Eventually, petroleum import concerns and global climate change will become more urgent—the result of increasing oil imports by the United States, rapidly expanding oil consumption in developing countries leading to probable increases in the volatility and level of oil prices, and increasing concentrations of GHGs. But calls for reduced petroleum use, GHGs, and pollution run up against the strong desire for travel. The two fundamental strategies to reduce energy use and pollution are (*a*) reduced travel (by encouraging transit use, less driving, and other shifts away from car use), and (*b*) the introduction of more benign vehicles and fuels. Below is a discussion of vehicles and fuels, a more likely and potentially far more effective strategy.

NATURAL GAS AND ALCOHOL FUELS

Enthusiasm for alternative transportation fuels has waxed and waned since the turn of the century, in the United States and elsewhere (Sperling 1988). The national Energy Policy Act of 1992 set a goal of 10 percent market penetration by alternative transportation fuels by 2000 and 30 percent by 2010; to initiate the transition, it adopted ambitious requirements for alternative-fuel vehicle purchases by government and private fleets, amounting to millions of vehicles by 2000.

Despite the existence of strong economic and environmental constituencies for alternative fuels, and strong rhetoric on their behalf by the federal government and some states, their penetration has been slow. As of 1995, less than 2 percent of vehicle fuel consumption might be characterized as alternative fuels. Less than 1 percent is ethanol made from corn, fewer than 0.2 percent of vehicles (250,000) are powered by propane, about 50,000 vehicles are powered by natural gas (virtually all are bi-fuel with gasoline), about 15,000 are methanol- and ethanol-compatible (but rarely operate on the alternative fuel), and a few thousand are powered by electricity (ORNL 1995). The corn-ethanol fuel is twice as expensive as gasoline and exists only because of a generous federal tax subsidy. Propane vehicles have existed for decades, but propane is derived from petroleum and natural gas and is in limited supply. Methanol is no longer considered a viable alternative for ICE vehicles, because other options are economically and environmentally superior, though one day it may gain acceptance as a fuel for fuel cell vehicles.

Natural gas has some promise. It is less expensive than gasoline, accounts for about 20 percent fewer GHGs (on a fuel cycle basis) when used in a vehicle dedicated (and optimized) to that fuel (Deluchi 1991), and emits considerably fewer emissions of conventional pollutants. The potential environmental benefits are considerably smaller than those from electric-drive options, and the refueling infrastructure is expensive (at least $300,000 additional cost per fuel station). The major automakers are offering light-duty natural gas vehicles (NGVs) for sale, but future sales are uncertain, especially in the light of a 1995 decision by the national Natural Gas Vehicle Coalition, the premier advocacy group for NGVs, to focus its support on heavy-duty applications, where the potential air quality benefits are much larger (both particulates and nitrogen oxides can be sharply reduced) and the fuel consumption per vehicle is much higher (leading to greater economies in refueling infrastructure).

The most effective alternative fuel for reducing GHG emissions is alcohol made from cellulosic biomass. The preferred biomass feedstock for transport fuels currently is corn (and sugar cane in Brazil). This preference for corn is due to large federal and state subsidies, and the ease of producing the ethanol fuel. But the cost is twice that of gasoline, and the air pollutant and greenhouse benefits are negligible. A more attractive biomass fuel option, from an energy and environmental perspective, is to convert more abundant cellulosic material—such as trees, grasses, and solid waste—into ethanol or methanol. But the cellulose conversion processes are more complex and not commercially proven, the capital costs greater, and lead times longer because of the need to create energy plantations around the site of the process plant (because feedstock collection and transport are expensive). Thus, even though production costs are expected eventually to be much less than corn-ethanol costs, and even though cellulosic biomass fuels are eligible for the substantial subsidies now received by corn-ethanol, there has been no significant investment in the United States in cellulosic biomass fuels for use in transportation. The National Renewable Energy Laboratory continues to receive about $30 million per year for research and development (R&D) on improved cellulosic conversion processes.

The 10 and 30 percent alternative-fuel goals of the Energy Policy Act of 1992 are not likely to be attained. The centerpiece of the act's strategy for reaching those goals is a set of rules requiring fleets to switch to alternative fuels. But in the 4 years since the act was passed, only the rules pertaining to federal fleets had been adopted, and funding for that program had been mostly eliminated. Rules for state government and fuel provider fleets are pending, and rules pertaining to the vast majority of fleet vehicles, those belonging to nonenergy businesses, are not under consideration at this time. Where fleet rules *are* adopted, the preferred choices are fuel-flexible alcohol vehicles (which are almost always fueled with gasoline) and NGVs, but congressional enthusiasm for fleet rules has mostly dissipated.

ELECTRIC-DRIVE TECHNOLOGY

EVs encompass a much wider range of technologies than just battery-powered vehicles, and the potential benefits are far broader than air quality. One can hybridize a small ICE (e.g., gas turbine, gasoline, or diesel engine) with an electric motor, by combining it with a small energy storage device such as a flywheel, ultracapacitor, or small battery. Alternatively, a fuel cell could be substituted for the ICE. The advantages of electric-drive vehicles are many, but they vary depending on the source of energy and the combination of power system technologies. It is this rich profusion of technological opportunities that makes electric-drive vehicles so attractive. In addition to energy and environmental benefits, various technological combinations provide consumers with the benefits of less noise, lower energy cost, greater reliability, longer vehicle life, less maintenance, and the ability to recharge at home (Sperling 1995).

Although these various attractions exist, they have not been sufficient to inspire automotive companies to invest seriously in electric-drive vehicle technology. The start-up costs and risks are too large. California's ZEV mandate, premised solely on air quality benefits, has been the principal motivation. But electric-drive vehicles provide other large nonmarket benefits: reduced use of petroleum and GHGs, in virtually all combinations and settings. These reductions, approaching 100 percent for some combinations and fuels, are the result of greater energy efficiencies with electric drive and the greater potential for fuel substitution. The point is, electric-drive vehicles provide the potential for huge improvements along a number of environmental dimensions, not just air quality. If the momentum is to be sustained by government action outside California, it will have to be for reasons other than just air quality.

But all environmental effects of electric-drive technology are not uniformly positive across technologies and space, which creates even more uncertainty over government support for electrics. For instance, the emissions benefits of battery EVs are much greater in regions with very clean electricity generation, such as California, than in those that burn mostly coal. The use of batteries introduces large amounts of new materials into the environment, some of which may be toxic. This problem, seepage of battery materials into the environment, may be more perceptual than real, however. The more toxic materials, such as cadmium, are likely to be restricted, and others are likely to be almost completely recycled.

Very little of the lead from the more than 70 million lead-acid batteries sold each year for ICE vehicles ever causes a health risk because virtually all the lead is recycled, and lead processing plants are tightly controlled. (Lead levels in blood dropped 86 percent between 1960 and 1990, from 20 μg/dL to 2.8, despite increased sales of lead-acid batteries and a 17 percent increase in total lead usage per capita over that period. The drop in lead levels was due to reduced use of lead in gasoline. Industrial production and the use of lead in batteries are considered a minor health threat.) In any case, lead-acid batteries are unlikely to gain much usage in EVs, and other battery materials are likely to be less toxic. Moreover, the large size and weight of batteries and the high value of the materials almost ensure close to 100 percent recycling of traction batteries. The reality is that all new technologies and fuels will have some adverse environmental consequence; the regulatory process will guide investment choices toward those choices that are more benign.

The roadblocks to electric-drive vehicles are many. These roadblocks have much to do with uncertainty over cost and performance, as well as public commitment to energy and environmental goals. If costs and performance do not improve sufficiently and government fails to reward the energy and environmental advantages of more benign vehicles and fuels, the market for electric-drive vehicles will be limited to niches. Technological progress and government intervention are a function of many factors, most of them linked to corporate and consumer support.

How the interplay of interest groups will play out is difficult to determine. Certainly, the powerful oil industry, whose economic interests are threatened, will oppose (and have opposed) government support for electric-drive vehicles, but other industrial interests will provide support, including the electricity and perhaps natural gas industries, and the various high-technology industries that see an opportunity to expand their sales to the automotive industry.

Many believe that EVs are doomed to failure, at least in the near future, because of automaker skepticism. Automobile industry response is complex and difficult to predict. But unlike oil companies, which see no self-interest in battery electric vehicles, and relatively little in other electric-drive options, automakers are less unified. Automakers oppose mandates as a matter of principle, but they have become a highly competitive global industry in the past few decades, and many companies are eager to gain an edge over their competitors. The likelihood of a large market for electric-drive vehicles outside the United States provides extra motivation to automakers to sustain investments in electric technology. The automaker response to the California regulators and California market, as well as the potentially larger international market, cannot be predicted. It appears likely that battery EVs will continue to be pursued; the question is how broadly and successfully; likewise, with hybrid and fuel cell technology.

SMALL VEHICLES

One ancillary outcome of government support for EVs could be the creation of a market in the United States for very small cars and trucks. Such vehicles are common elsewhere: about a fourth of vehicles sold in Japan are minivehicles (defined as having engines of less than 660cc, about half the size of the smallest engines sold in the United States). Automakers are responding to the high cost and large size of batteries by building hybrid and fuel cell vehicles, but another response is to build smaller vehicles. Very small vehicles of the type sold in Japan and elsewhere would require only very small battery packs, which would not have the cost or weight burdens of larger cars and trucks.

Already entrepreneurial companies are exploring the market for a variety of very small vehicles, from electric-assist bicycles for difficult terrain and less fit individuals; to "glorified" golf carts sold by some companies in retirement and resort communities, national parks, and large industrial, military, and educational campuses; to somewhat larger vehicles for nonfreeway use. These vehicles have a variety of attractions: they are generally easier to operate and park, cost less, are rechargeable at home, require less space, consume less energy, and produce less pollution. Small gasoline-powered vehicles could be designed to give some of the same benefits as small EVs, but the much more positive image of electrics, government incentives for their purchase and use, and the relatively small cost handicap of small battery packs mean that most small vehicles are likely to be battery-powered electrics.

IMPLICATIONS FOR THE TRANSPORTATION COMMUNITY

The transportation community confronts four sets of responsibilities in responding to the prospects for more benign fuels and vehicles: R&D, infrastructure design and investments, regulatory policy, and financing.

Transportation agencies have sponsored little research on advanced vehicle propulsion technology and nonpetroleum fuels. The most notable program in this regard was the Federal Transit Adminstration's funding of a fuel cell bus program. A review of the PNGV program by a National Research Council committee criticized the U.S. Department of Transportation (DOT) for allocating no funds to PNGV technologies (NRC 1996). Virtually all federal vehicle propulsion R&D is funded by DOE. DOT has practically no expertise in fuels and electric-drive vehicles. Its primary interest has been in helping transit operators accommodate alternative fuels and in worrying about how to collect fuel taxes in a post-gasoline era. This attitude of benign neglect could continue, but at considerable risk.

The risks are the following. First, other federal and state programs will direct vehicle technology and fuels in directions that are not compatible with road infrastructure programs and deployment plans for intelligent technologies. For instance, electronics capabilities of EVs are very different from those of gasoline vehicles. Likewise, determinations of air quality conformity could be sensitive to vehicle propulsion and fuel strategies. And emergency services may be different for EVs.

Second, opportunities to diversify vehicle technology are stunted by rules and regulations aimed at standardizing all streets and highways for use by all vehicles, and safety standards that ignore the entire road system in favor of highly specific and uniform vehicle standards.

Third, the federal system for regulating fuels and vehicles is rigid and is tied to existing technology. It is not well suited to the different emissions, energy, and safety attributes of new fuels and vehicles; it hinders innovation, does not allow trade-off between different attributes, and is insensitive to regional differences. While the Environmental Protection Agency plays a lead role in most of these issues, DOT and the Federal Highway Administration (FHWA) have an abiding interest.

Fourth, the transition away from gasoline and diesel fuel can be seen as a threat to the financial integrity of the transportation financing system, or as an opportunity. FHWA could passively await steadily diminishing gas tax revenues or could start devising new methods that are more rational and equitable (Reno and Stowers 1995).

CONCLUSIONS

In summary, abetting the public pressure to create more benign vehicles and fuels is a far-reaching revolution in various vehicle-related technologies. Recent and continuing advances in storing electricity and gases, electrochemically converting chemical fuels to electricity, biologically converting cellulose to chemical fuels, designing less expensive and bulky electronics, storing and manipulating information, and manufacturing inexpensive lightweight materials are bringing more benign vehicles closer to commercial reality. What is unknown, and unknowable, is which of these technological improvements will be commercialized first and in what combinations. The implications for the transportation sector are not revolutionary, but they could be significant and far-reaching.

REFERENCES

Abbreviations

NRC National Research Council
ORNL Oak Ridge National Laboratory

Deluchi, M.A. 1991. *Emissions of Greenhouse Gases from the Use of Transportation Fuels and Electricity*. Report ANL/ESD/TM-22, Vol. 1. National Technical Information Service, Springfield, Va.

Greene, D.L. 1996. *Transportation and Energy*. Eno Foundation, Inc., Lansdowne, Va.

Heavenrich, R.M., and K.H. Hellman. 1996. *Light Duty Automotive Technology and Fuel Economy Trends Through 1996*. U.S. Environmental Protection Agency. Report EPA/AA/TDSG/96-01.

Mintz, M., and Vyas, A. 1993. Why is Energy Use Rising in the Freight Sector? In *Transportation and Global Climate Change* (D.L. Greene and D.J. Santini, eds.), ACEEE, Berkeley, Calif.

NRC. 1996. *Review of the Research Program of the Partnership for a New Generation of Vehicles.* Washington, D.C.

ORNL. 1995. *Transportation Energy Data Book*, 15th ed. Report ORNL-6856. National Technical Information Service, Springfield, Va.

Reno, A.T., and Stowers, J.R. 1995. *NCHRP Report 377: Alternatives to Motor Fuel Taxes for Financing Surface Transportation Improvements.* TRB, National Research Council, Washington, D.C.

Sperling, D. 1988. *New Transportation Fuels: A Strategic Approach to Technological Change.* University of California Press, Berkeley.

Sperling, D. 1995. *Future Drive: Electric Vehicles and Sustainable Transportation.* Island Press, Washington, D.C.

Appendix F

Impacts of the Highway Transportation System on Delivery of Health Services and Social Services

Mark Baldassare
University of California, Irvine

The ability of residents to access health and social services is critical to the well-being of the U.S. population. All of us are in need of medical care on a regular basis; some residents depend on publicly provided health and social services for their daily survival. This paper explores the ways in which the current highway transportation system influences access to health care. In order to consider the potential for change in the future, it is important to understand that demographic trends are converging to have an impact on both the transportation system and the delivery of health and social services. At the same time, we need to understand that both transportation and health and social services are in transition.

We make several basic assumptions in exploring the relationships between the highway transportation system and health and social services. First, the accessibility of services is by nature a challenging task in the metropolitan society that we live in today. This is because most U. S. residents live in large and geographically sprawling suburban regions far from the downtown business districts, where public and private services were once concentrated. A diverse array of needed services and facilities are thus spread out over a large area, often at great distances from residential areas. Next, mobility problems for "vulnerable populations" present even greater hurdles for the groups that need health and social services the most. For instance, the inner-city poor, the elderly, the physically or mentally disabled, recent immigrants, and single parents with young children often lack the money, mode of transport, time, or personal assistance needed to travel long distances to obtain health and social services. Finally, the supply of health and social services may not be keeping pace with the demand. There are pressures today in the private sector to reduce health care costs, and the public sector is seeking ways to reduce health and welfare expenditures. Thus, fiscal realities are also affecting the public's access to social and health services.

DEMOGRAPHICS

Several powerful demographic trends in the United States today are shaping both personal travel and the demand for health and social services. They include suburbanization, economic restructuring, foreign immigration, aging of the population, changing family roles and household composition, and the persistence of an urban underclass.

The U.S. population has been undergoing a dramatic "urban deconcentration," or suburbanization, for 50 years. Large and sprawling metropolitan areas have replaced central cities as the dominant urban form in every region of the nation (Baldassare 1992). There are more than 300 metropolitan areas, and they all contain a predominantly suburban population. Almost half of the approximately 250 million population in the United States now lives in the suburbs. In fact, most of the population growth in recent decades has been in suburban locales. For instance, 86 percent of U.S. population growth since 1970 has gone to the suburbs, while 76 percent of metropolitan growth in the 1980s has been in the suburbs (FHWA 1995). Employment has also grown at an explosive rate in the suburbs and is expected to show continued growth in the future. For instance, between 1960 and 1980, suburban employment grew from 14 million to 33 million jobs, a rate of 136 percent (Palen 1995).

The suburbanization of the United States has had profound sociological implications. Many central cities have experienced the loss of middle-class residents, higher-paying jobs, and tax revenues to the suburban ring of the metropolis. The rural areas bordering the metropolis are confronted with development pressures at the expanding outer rings of the suburbs. Meanwhile, the suburban communities themselves are being transformed from low-density, residential, and socially homogeneous locales to large, sprawling, and socially diverse suburban regions. The urbanization of suburban regions has resulted in the emergence of community conflict over growth, "NIMBY" (i.e., not in my backyard) attitudes expressed toward the location of public and private facilities, and local problems such as traffic congestion, school overcrowding, crime, and racial tensions (Baldassare 1992; Baldassare 1994).

Another important trend is economic restructuring. U.S. society is being transformed into a "post-industrial" economy in which manufacturing jobs are on the decline and service employment is becoming dominant. At the same time, the central cities have experienced job decline and the suburbs have benefited from the relocation of urban companies and the growth of new industries. Recently, large companies have gone through a period of mergers, globalization, computerization, and downsizing that has created massive layoffs from large companies. These trends result in a labor force that is becoming more oriented toward high-technology manufacturing, service employment, smaller companies, self-employment, and suburban locales (FHWA 1995).

The economic restructuring under way has profound implications for U.S. society. The loss of manufacturing employment has resulted in the displacement of skilled blue-collar workers, whereas the downsizing of large corporations has affected middle-class professionals. The growth of suburban employment has helped many in the middle class, but the loss of inner-city jobs has been detrimental to the urban poor (Sassen 1990). The growing importance of the service sector results in a mixture of many low-paying and high-paying jobs. New technical skills and high educational standards are required of many displaced workers who are seeking new employment. The trend toward smaller and more scattered workplaces strains public and private services for workers.

Next, foreign immigration from Asia and Latin America is dramatically reshaping U.S. society. The Asian population more than doubled and the Hispanic population increased by more than half in the 1980s (Frey 1993). In the meantime, the non-Hispanic white population, which has experienced a lower rate of foreign immigration, has been increasing at a much slower rate. The result of foreign immigration is a nation that is becoming more racially and ethnically diverse. The 1990 Census reports that about 60 million Americans are in minority groups, including African American (30 million), Hispanic (22 million), and Asian (7 million). The U.S. population is estimated to be 15 percent African American, 11 percent Hispanic, and 4 percent Asian. The minority population will make up an even greater share of the U.S. population in the future (FHWA 1995).

An increase in the minority population, particularly because it is growing rapidly as a result of foreign immigration, is having significant impacts on U.S. society. Foreign immigrants from many destinations are bringing to the

United States a host of new cultures, languages, values, consumer patterns, and expectations. This sometimes causes racial conflict and ethnic tensions between Asians, African Americans, Hispanics, and non-Hispanic whites. In an extreme form, this was evident in the 1992 Los Angeles riots (Baldassare 1994). Foreign immigrants also tend to be young adults with children. This places burdens on public institutions such as schools. Some are poor or displaced peoples who require government programs and private assistance. The effects of the minority population increase have been felt in many metropolitan areas, but most notably on the West Coast as a result of both Hispanic and Asian immigration. In a dramatic reversal from past trends, however, suburban areas as well as cities are experiencing a rapid growth in the Asian, Hispanic, and African American population. This is because both new immigrants and minority city dwellers are seeking the jobs and housing of the outer metropolitan ring. The combination of multiracial and ethnic change, and a rapid rate of foreign immigration, has placed special strains on the people and institutions in the suburbs (Baldassare 1992; Baldassare 1994).

Another important trend is the aging of the U.S. population as a result of medical advances that extend life expectancy. In 1990 more than a fourth of the U.S. population was over age 60. Older Americans now represent the fastest-growing age group. In fact, statistics indicate a very rapid increase in the growth of the "old" old—that is, Americans over 75 years of age (FHWA 1995). Policy makers are alarmed about the future numbers and rate of growth of the older population. The fact that the leading edge of the baby boom generation is now turning 50 means that in a decade the nation will begin to feel the effects of a historically large age cohort reaching old age. That effect will be long-lasting because the baby boom occurred for about two decades, and it could be even more dramatic if medical advances continue to extend the average life span.

The growth of the older population is having sizable effects on both public and private services. Much in the same way that the original baby boom placed high demand on schools and housing for young children, the aging of the population puts enormous pressures on the American institutions responsible for caring for older adults. Another important fact is that the older population is a geographically dispersed and socially diverse population. Most live in the suburbs of metropolitan areas, while many are spending their retirement years in rural communities. The older population is also showing greater ethnic and racial diversity, as well as a range of economic conditions (FHWA 1995). These trends add challenges to providing services for older Americans.

Next, dramatic changes in family roles and household composition are occurring in the United States. Women have entered the labor force in record numbers during the past 20 years. This reflects a shift from the "traditional" family roles in which men worked outside the home and women stayed at home with the children. According to the 1990 Census, 6 in 10 married women are now employed outside the home. Half of women with younger children are working, and three in four of those with children under 6 years of age are employed. Other important changes involve divorce and adolescent pregnancy. Both are contributing to a growth in single-parent households headed by women and a large number of employed and unemployed women with children who are in poverty (FHWA 1995).

The changing patterns of family and work roles are placing special burdens on individuals and institutions. Working mothers need to divide their time and attention among the demands of the home, workplace, and providers of child care. Working fathers are called on to take a greater role with home and child care. Public institutions geared toward children, including schools, must cope with the fact that fathers and mothers are absent from the home. The delivery of child care services, by public and private providers outside the family, has taken on a more critical role.

Finally, the persistence of an "urban underclass" is an important trend in U.S. society. Sociologists have identified the existence of a sizable group of inner-city minorities that show a pattern of extreme poverty, chronic unemployment, and segregation, as well as daily struggles with crime, violence, broken families, drug and alcohol abuse, poor schools, health problems, and inadequate housing (Wilson 1987). The urban underclass is made up mostly of African Americans, but it increasingly includes Hispanic immigrants. An estimated 4 million poor African Americans and Hispanics live in urban poverty areas, mostly in big cities such as New York, Los Angeles, Chicago, and Detroit (Kasarda 1993).

The persistence of an urban underclass in an era during which civil rights laws were supposed to help eliminate minority poverty is a frustrating reality with dire consequences. The outlawing of housing and employment discrimination was seen as opening new opportunities for all inner-city minorities. But many inner-city Hispanics and African Americans remain isolated from suburban housing and high-paying jobs (Massey and Denton 1993). Some say that ongoing trends in society—such as suburbanization, economic restructuring, and foreign immigration—have worsened the plight of the inner-city poor. In any event, the urban underclass today is highly dependent on public assistance programs for its daily survival and future hopes.

HEALTH AND SOCIAL SERVICES

Major changes are occurring in the delivery of health and social services. The demographic trends that we have reviewed are also playing roles in the accessibility of health and social services.

Today, private health care providers find themselves under pressure from the government, employers, and insurers to contain the rising costs of providing medical care. The concerns with health care costs are (*a*) a response to rising Medicare expenditures as the population of older Americans has expanded, (*b*) a reaction to the increasing availability of expensive medical procedures, and (*c*) an effort to contain health care inflation. The response has been to offer health services through "managed care" and health maintenance organizations, in which patients have fewer choices of health care providers and more limited access to medical treatment. Thus, there is decreasing emphasis on extensive hospital stays and inpatient procedures and a greater orientation toward outpatient facilities and in-home care. The overall trend in cost containment means that medical patients will spend less time in more different kinds of health care settings.

Many in the public sector are calling for the federal government to play a lesser role in providing health care for those who cannot afford it, or social services for those who cannot manage their own affairs. Some argue that health and social service costs need to be shifted away from the federal government to allow the state and local governments more authority. Others are calling for local charities, churches, and private agencies to have increasing responsibilities in delivering health and social services. The proposals for decreased federal expenditures in health and social services are a result of fiscal pressures, including the federal debt, growing expectations for a balanced federal budget, and voters' reluctance to pay higher taxes. The result is, once again, less access to services and the diffusion of service locations.

Another new trend involves the kinds of health and social services that the public and private sectors are providing. Public health and social programs are being criticized as costly, inefficient, and sometimes even counterproductive to their clients. There is a growing consensus that the "urban underclass" remains in desperate conditions even after receiving government assistance, and concerns that groups such as immigrants, the homeless, and single parents with young children are too dependent on government support. The result is a growing interest in providing a wider range of programs. In health care, some are proposing that prevention programs be expanded to avoid the high costs of providing medical treatment. Some are calling for "workfare" programs to replace welfare cash payments. Others say that more education and training for the unemployed are needed, or point to the importance of providing child care for working single parents and unemployed welfare recipients with young children. These new programs, taken together, point to the likelihood that health and social services will become more decentralized and geographically dispersed in the future. This, again, raises issues about their accessibility.

Major demographic trends are placing demands on the delivery of health and social services. Suburbanization has resulted in the deconcentration of health and social services over a large geographic area, thus creating challenges for its accessibility. In addition, the fact that suburban regions are politically fragmented places where local government is decentralized will hinder efforts to shift the responsibility for public services from the federal to the local level. Economic restructuring has caused massive layoffs from large companies and a variety of needs for health and social services by those who have become unemployed. The aging of the population means that more Americans require medical and social services that they can reach easily on a regular basis. We can also anticipate a huge increase in the demand for medical services, and escalating Medicare costs, when the baby boomers reach old age in a decade. The growing numbers of working women with children and poor single mothers have increased the need for child care and children's health facilities that are convenient to home and work. The immigrant population is raising the demand for children's health and social services, and it remains a leading issue in the ongoing debate about who should receive public services. The urban underclass lacks the knowledge and the means to access the health and social services that are spread out over a large metropolitan region, and their presence fuels the controversy over the kinds of programs that can make a difference.

IMPACTS OF THE HIGHWAY SYSTEM

The public's reliance on the highway system for metropolitan travel has important consequences. The impacts are evident for all services, and even more noticeable in the special case of delivering health and social services. Further, the need for private automobiles and highways to reach health and social services creates special problems for demographic groups with limited mobility. In addition, highway crashes have direct effects on health and emergency services.

It is important to note that urban deconcentration has placed enormous demands for commuting on the highway system. Today, the most common commuting pattern is travel from a suburban home to a suburban workplace. The rise in suburban commuting means that more Americans are making longer work trips, often on highways, and that fewer travel to central business districts (CBDs) (FHWA 1995). One implication is that traffic congestion is spread over a larger region. Another is that public transit use and ridesharing are impractical travel options for commuters from suburban homes to workplaces. Driving alone in a private automobile is thus by far the popular choice for the suburb-to-suburb commuters.

Suburbanization has also created greater demands on the highway system for residents who are seeking services. Public as well as private facilities are deconcentrated in metropolitan regions. This means that the services that people need are spread over a large geographic area. Similar to commuting trends, travel usually involves individuals driving themselves in their private automobiles from their homes to their service destinations. There is little use for public transit because few facilities are located in the downtown business district.

Highways play a central role when residents need health and social services. These facilities are also found in moderate-density areas and spread out over large geographic areas. Travel to health services can cause problems when people are sick or have other physical problems that make it difficult for them to drive themselves frequently over long distances to multiple settings. Since commuting trends mean that suburban "rush hours" can go in all directions, over extended periods of the day, they can also lengthen the time that it takes to travel to and from health and social services. In sum, the combination of suburban locations and private automobile use can cause difficulties for people who are seeking health and social services.

The important role of highway travel in the delivery of health and social services is even more critical when we consider residents who have limited access to private automobiles. Many of the demographic trends reviewed earlier point to groups who have chronic problems with mobility. Older Americans often do not drive or cannot drive because of physical disabilities. Foreign immigrants can lack the knowledge needed to use the highway system effectively. Single parents with young children may not be able to manage a lengthy drive over long distances because of commitments at home and at work. The urban underclass is highly isolated and very dependent on public transit and local institutions to meet its needs. Thus, all of these "vulnerable" groups can have more difficulty than the average resident in reaching health and social services.

Highways also have a direct impact on the demand for health services through motor vehicle crashes. A report by the Public Health Service (PHS 1993) indicates that there were 5.4 million nonfatal injuries and 44,531 fatalities related to motor vehicle crashes in 1990. They estimated $13.9 billion in direct medical care expenses for these crashes. Another study (FHWA 1991) also found that 5 million people were injured in motor vehicle crashes and 47,000 died as a result of these accidents in 1988. Medical costs in this study were estimated at $11 billion and included hospitals, doctors, pharmaceuticals, equipment, institutional care, and home care. Emergency costs, including police, fire, and ambulance service, are a relatively minor part of the costs of highway crashes. FHWA also reports that high-speed crashes, such as those that occur on highways, are the most costly because they often involve serious injuries and property damage. It is estimated that employers pay 20 percent and the general public pay 48 percent of the out-of-pocket costs (e.g., property, medical, emergency, legal, and productivity costs); accident victims and their families pay the rest, including pain and suffering and losses in productivity. Obviously, highway crashes have a sizable effect on the use of health services.

A recent report on highway safety found that highway fatalities are on the decline (FHWA 1995). Measured in terms of the numbers of persons injured per 100 million vehicle-mi traveled, there was a significant decrease in fatalities between 1983 and 1993. In general, the fatality rates are lower for urban than for rural highways, and for Interstates than for other types of highways. This study points out that highway condition and performance affect accident and fatality rates. Other factors include driver performance, weather, law enforcement, and automobile safety.

FUTURE IMPACTS OF THE HIGHWAY SYSTEM

In this section, we speculate about the future impacts of the highway system on health and social services. We base assumptions on current travel patterns, demographic trends, and the delivery of health and social services. We also consider future possibilities if there are dramatic changes in today's trends.

The highway system has provided the infrastructure that led to urban deconcentration and suburbanization. This physical reality of metropolitan regions will not change anytime soon. Today, a variety of services, including health and social services, are spread over large geographical areas within the metropolis. Clients of services often need to travel long distances, to multiple locations, in order to receive the needed services. For health services, managed care and the emphasis on outpatient care add to the number of trips, time, and distance. In social services, the movement toward providing training and child care to encourage outside employment also adds to the numbers of trips, time, and distance. A continuation of trends points to problems in access to health and social services for the general population in the future.

Of course, large numbers of Americans are less geographically mobile. They lack access to private automobiles or have difficulties in driving alone on highways for long distances. Because of travel demands, they have more problems in reaching health and social services. Most experts anticipate a growth of older Americans and minority immigrants, and no one is predicting fewer poor single parents with young children or members of the urban underclass. In the future, there may be more members of "vulnerable populations" who will have trouble reaching the health and social services they need.

The accessibility of services is only one of the concerns raised by increasing numbers of solo drivers using highways to reach health and social services. Other important issues include highway safety and congestion. An increase in the numbers of people driving alone on highways to reach health and social services may ultimately lead to more accidents and more traffic congestion. This could result in more demand for health services, emergency health services, and highways. Moreover, it should be considered that there will be a significant growth in the older population beginning in a decade. The older population tends to be more accident prone, and it is more likely to travel often for health and social services. The growth of the older population could add to accidents and congestion on the highways.

Several factors, were they to change, would further accentuate the general trends of increased inaccessibility, and the more difficult circumstances for vulnerable groups, that we are predicting.

First, the federal government could reduce drastically its funding for highways and public transit. This could greatly worsen the issue of access to health and social services by the general population. It would be especially crippling for vulnerable groups, who are heavy users of public buses for travel to services. In addition, if funds for highway safety were cut and accidents rose, this would increase the demand for health and emergency services.

Next, the federal government could reduce drastically its funding for health and social services. Moreover, state and local governments and private sources may not have the financial resources and political support to administer these programs properly. Thus, in the future, there could be larger numbers of people seeking more limited health and social services. Consider the possibility of great numbers of people going to only a few locations to receive health and social services. This would obviously place enormous strains on existing services and on the highway system leading to them.

Finally, we may be seriously underestimating the growth of demographic groups that both use health and social services and have difficulty accessing these services. For instance, medical advances in the next century could lead to longer life expectancy. In this case, we could be drastically underestimating the numbers of older Americans, and especially the "old" old, who will be using the highway system to seek health services. Another possibility is that foreign immigration could remain at current high levels, or actually increase, as a result of globalization of the economy or international strife. If we are underestimating the future impacts of immigration, then we have not fully accounted for the growing diversity of the population and the highway system's impacts on health and social services.

RESEARCH ISSUES

This paper has reviewed the accessibility of health and social services, which is a challenge in a metropolitan society. It has pointed out that some groups with limited mobility face special difficulties when they are trying to reach the health and social services that they need. Further, the combination of demographic trends and fiscal realities suggests that the issue of accessibility of health and social services, for the general population but especially vulnerable populations, will probably increase over time.

It is important to note that the groups with mobility problems are often a major audience for health and social services. Special efforts are required to make sure that these groups receive the services they need. One approach is to provide door-to-door taxi service for these special groups. Another is to localize health and social services in satellite facilities near their homes. This would include, for instance, health facilities near retirement communities and social services near immigrant areas and the urban poor. The localization of services reduces the likelihood of citizens' having to drive long distances or not receiving care. An alternative is to centralize the health and social services in the areas that are better suited to public transit or better served by highways. This latter approach requires a new level of cooperation among local governments to enable them to coordinate their land use, health and social services, and transportation planning at a regional level.

Another recommendation is to improve public information about the location of health and social services and travel to these services. As a result of their geographic dispersal, many health and social services may be hidden from the populations that serve them. The inner-city underclass is but one example of people who are cut off from knowing where needed services are and how to get to them. Recent immigrants with limited language skills are another case. As the older population grows, information on how to reach health and social services through a fast and congestion-free highway route could become as critical as how to travel to and from work. Thus, there is a potentially important role for telecommunications and computers to improve access and travel to health and social services.

There are several suggestions for future research and development on the impacts of highways on health and social services.

First, we need to know more about the impacts of suburbanization on the use of highways and the demands for health and emergency services. For instance, it was recently reported that suburbs are more dangerous places than central cities. This is because the greater use of automobiles in suburbs, and thus the increased likelihood of traffic crashes, far outweighs the risks of crime victimization in cities (Gerstenzag 1995). We need thorough research on the relationships among suburban daily travel, traffic crashes, and the need for health services and emergency health services.

Next, we need more detailed studies of personal travel to health and social services. A great deal is known from the U.S. Census and other information sources about how Americans commute from home to work. But, increasingly, we have come to recognize that commuting is one of many purposes that highways have in a suburban society. There is little systematic information, for instance, on American's travel to health facilities, physicians, child care centers, and social service agencies. It would be important to know the frequency of visits, number of sites, distances, time, means of travel, and use of highways. Only then could we have a thorough analysis of travel and usage patterns of the general population and subgroups. This is critical to understanding current and future demands on the highway system.

Furthermore, detailed analyses of the travel demands being created by current demographic changes are needed. For instance, there may be many impacts of having a large and growing portion of the population older that 75 that needs to have access to services in a metropolitan region. The trend of providing child care for poor working mothers may affect traffic congestion in residential areas. Another possibility is that large numbers of immigrants could begin, at some point, to drive alone to health and social services. Economic restructuring and self-employment may result in less rush hour traffic or a greater use of highways to reach services during work hours.

Finally, research is needed on how future demographic and service trends may affect highway safety. For instance, there may be more accidents as "old" older populations drive on highways at fast speeds to and from health service providers. There may be more accidents as immigrants with limited knowledge of the area and language are using highways that have English-only signs. There may be more accidents involving young children as working mothers rush their children from home to child care settings before they travel to work. This research is needed to help make sure that the current positive trend of reduced highway fatalities will continue.

REFERENCES

Abbreviations

FHWA Federal Highway Administration
PHS Public Health Service

Baldassare, M. 1992. Suburban Communities. *Annual Review of Sociology,* Vol. 18, pp. 475–494.

Baldassare, M. 1994. *The Los Angeles Riots: Lessons for the Urban Future.* Westview Press, Boulder, Colo.

FHWA. 1991. *The Costs of Highway Crashes.* U.S. Department of Transportation.

FHWA. 1995. *1995 Status of the Nation's Surface Transportation System: Condition and Performance.* U.S. Department of Transportation.

Frey, W. 1993. People in Places: Demographic Trends in Urban America. In *Rediscovering Urban America* (J. Sommer and D.A. Hicks, eds.), U.S. Department of Housing and Urban Development, pp. 3-1–3-106.

Gerstenzag, J. 1995. Cars Make Suburbs Riskier Than Cities, Study Says. *Los Angeles Times,* April 15, p. 1.

Kasarda, J.D. 1993. Inner-City Poverty and Economic Access. In *Rediscovering Urban America* (J. Sommer and D. A. Hicks, eds.), U.S. Department of Housing and Urban Development, pp. 4-1–4-60.

Massey, D. S., and N. A. Denton. 1993. *American Apartheid: Segregation and the Making of the Underclass.* Harvard University Press, Cambridge, Mass.

Palen, J. 1995. *The Suburbs*. McGraw-Hill, New York.

PHS. 1993. Economic Impact of Motor Vehicle Crashes—United States, 1990. *Morbidity and Mortality Weekly Report*, Vol. 42, No. 23, pp. 443–447.

Sassen, S. 1990. Economic Restructuring and the American City. *Annual Review of Sociology*, Vol. 16, pp. 465–490.

Wilson, W. J. 1987. *The Truly Disadvantaged: The Inner City, The Underclass and Public Policy.* University of Chicago Press.

Study Committee Biographical Information

RAYMOND F. DECKER, *Chairman,* is Chairman of USP Holdings, Inc., and Thixomat, Inc., companies that commercialize new technologies. Dr. Decker has also served as Vice President of Corporate Technology for Inco, Ltd., and as Vice President for Research and Corporate Relations of Michigan Technological University. He was a member of the Strategic Highway Research Program (SHRP) Executive Committee and co-chair of the Transportation Research Board (TRB) SHRP Committee. Dr. Decker is a member of the National Academy of Engineering and is currently an Adjunct Professor at the University of Michigan.

DON C. KELLY, *Vice-Chairman,* is currently Vice President of Jordan, Jones & Goulding. Previously he served as the Secretary of Transportation for Kentucky. He also served 13 years with the Kentucky Department of Transportation in various aspects of engineering and planning and 2 years in the Kentucky Department of Commerce. From 1988 to 1991 he was a planning consultant with the Los Angeles Metro Rail Project. He has served on numerous advisory committees for TRB, the American Association of State Highway and Transportation Officials (AASHTO), and the state of Kentucky, and in 1995 was named by the American Public Works Association as one of the "Top Ten Public Works Administrators in the U.S."

ALLAN L. ABBOTT is currently Director and State Engineer of the Nebraska Department of Roads, having previously served with the Illinois Division of Highways from 1961 to 1991. He currently serves as Chairman of AASHTO's Standing Committee on Research and as Chairman of the TRB Long-Term Pavement Performance (LTPP) Committee. Mr. Abbott has a bachelor's degree from the University of Illinois and a master's from Sangamon State University.

RICHARD P. BRAUN, Special Consultant, founded the University of Minnesota's Center for Transportation Studies, serving as director from its inception until his retirement. He was the Commissioner of the Minnesota Department of Transportation from 1979 to 1986 and President of AASHTO in 1985. Previously he served more than 30 years in different posts with the Minnesota Department of Highways and was Chairman of the Metropolitan Airports Commission in Minneapolis. He also served as a Vice President and Principal Associate with Barton-Aschman Associates, Inc., directing projects in the Twin Cities Office. He served on the TRB Executive Committee and on the SHRP Executive Committee.

JOHN E. BREEN holds the Nasser I. Al-Rashid Chair in Civil Engineering at the University of Texas at Austin. From 1962 to 1985 he was the Director of the Phil M. Ferguson Structural Engineering Laboratory at the university. Elected to the National Academy of Engineering in 1976, he is an Honorary Member of the American Concrete Institute (ACI) and has served as chair of the ACI Building Code group. He is a fellow of the American Society of Civil Engineering (ASCE) and a member of the International Association for Bridge and Structural Engineering.

WILLIAM F. BUNDY is currently a Vice President with the Fleet Financial Group. He served as the Director of the Rhode Island Department of Transportation from January 1995 to September 1996. Prior to that he was a career submarine officer in the U.S. Navy. He has served as the President of the Northeast Association of State Transportation Officials, on the I-95 Corridor Coalition Board of Directors, on the TRB Executive Committee, and as Chair of the AASHTO Financial Management Subcommittee. Mr. Bundy is an Associate Participant in the National Automated Highway System.

A. RAY CHAMBERLAIN is Vice President of Freight Policy for the American Trucking Associations (ATA). Dr. Chamberlain served as the Executive Director of Colorado's Department of Transportation and as President of Colorado State University; he has also held several executive positions in the private sector. He has served as President of AASHTO and as Chairman of the TRB Executive Committee.

FORREST M. COUNCIL is the Director of the Highway Safety Research Center at the University of North Carolina. He is a past President of the National Child Passenger Safety Association and a member of the Association for the Advancement of Automotive Medicine. Dr. Council has twice received TRB's D. Grant Mickle Award for best paper in the area of operations, safety, and maintenance. He previously served on the TRB policy study panel entitled Committee for the Study of Relationships Between Vehicle Configurations and Highway Design and is currently a member of the Committee for Guidance on Setting and Enforcing Speed Limits.

HENRY (Hank) E. DITTMAR is the Director of the Surface Transportation Policy Project, a coalition of environmental, conservation, and planning groups that focuses on the policy implications of transportation programs. He has managed a bus line and an urban airport and has worked for the Metropolitan Transportation Commission in Oakland. He is active on several TRB committees, including the Committee on Intergovernmental Relations and Policy.

NANCY D. FITZROY is retired from the General Electric Company's Gas Turbine Division after working most of her career in Corporate Research and Development. A chemical engineer whose specialty is heat transfer, she was elected to the National Academy of Engineering in 1995. She is a Fellow and a past President of the American Society of Mechanical Engineers International and an Honorary Fellow of the Institution of Mechanical Engineers (UK). She holds honorary doctorates from the New Jersey Institute of Technology and Rensselaer Polytechnic Institute.

LARRY R. GOODE is the State Highway Administrator for the North Carolina Department of Transportation, where he has worked since 1972. Dr. Goode is also an adjunct professor of transportation at North Carolina State University. He serves on several committees of the Institute of Transportation Engineers (ITE) and AASHTO and is currently a member of TRB's Transportation Systems Management Committee. He chairs the AASHTO Special Committee on International Activities Coordination. He is a registered professional engineer in North Carolina and Virginia.

JEAN JACOBSON is the County Executive for Racine County, Wisconsin. She served on the Racine County Board of Supervisors from 1980 to 1995. A former Chair of the Transportation and Telecommunications Committee of the National Association of Counties (NACo), she is the current Vice Chair. She previously chaired the NACo Transportation Subcommittee. Ms. Jacobson is Secretary of the Southeastern Wisconsin Regional Planning Commission and is on the Board of Directors of the Racine County Economic Development Corporation.

JACK KAY is Chairman of the Strategic Planning Board of TransCore, an SAIC company; TransCore specializes in traffic control and engineering issues. He has served as an advisor to the World Bank on transportation planning and traffic engineering in many developing countries. Mr. Kay served as the Chair of the Board of Directors of ITS America, is a fellow of ITE, and chaired the ITE-IVHS Advisory Committee from 1990 to 1993.

NEVILLE A. PARKER is currently Herbert G. Kayser Professor of Civil Engineering at the City University of New York (CUNY) and Director, CUNY Institute for Transportation Systems. Dr. Parker has been active on TRB committees, including the Committee on Low-Volume Roads, the Committee on Transportation Education and Training, the Steering Committees for the 5th and the 6th International Conferences on Low-Volume Roads, the Subcommittee for Assessing Worldwide Low-Volume Roads Problems, and the Expert Task Group for LTPP Analysis.

GILBERT S. STAFFEND is a consultant in manufacturing operations, focusing on the introduction of new technology and organizational change. Previously he was Manager of Computer Integrated Manufacturing at Allied Signal for 7 years and a manager of Design, Planning, Development, and Operations for Ford for 18 years. He has served on the National Research Council's Computer Science and Telecommunications Board Committee to Study Information Technology in Manufacturing.

DALE F. STEIN is President Emeritus of Michigan Technological University, where he also served on the faculties of the Department of Metallurgical Engineering and the Department of Mining Engineering. He is past President of the Metallurgical Society of the American Institute of Mining, Metallurgical, and Petroleum Engineers and was named a fellow in 1979. Dr. Stein is also a fellow of the American Society of Metals. He was elected to the National Academy of Engineering in 1986.

C. MICHAEL WALTON is Professor of Civil Engineering and holds the Ernest H. Cockrell Centennial Chair in Engineering at the University of Texas at Austin. In addition, he has a joint academic appointment in the Lyndon B. Johnson School of Public Affairs. He is a founding member of ITS America, where he chairs the Technical Coordinating Council and serves on the Board of Directors. Dr. Walton is a fellow of ASCE and ITE, and a member of the National Academy of Engineering, the Institute for Operations Research and the Management Sciences, Urban Land Institute, Society of American Military Engineers, Society of Automotive Engineers, Council of University Transportation Centers, and National Society of Professional Engineers. He is a past chair of TRB's Executive Committee and on the Board of Directors of both the International Road Federation and the International Road Educational Foundation.

RICHARD P. WEAVER recently retired as Deputy Director and Chief Engineer for the California Department of Transportation, where he worked for 37 years in all aspects of highway and transportation engineering, agency administration, and management. Before retiring, Mr. Weaver was the Chairman of AASHTO's Highway Subcommittee on Traffic Engineering and a member of AASHTO's Standing Committee on Highways. He is currently a member of the TRB SHRP Committee and the Chairman of the TRB SHRP Subcommittee for the Workshop on Pavement Renewal for Urban Freeways. He is also a member of the TRB Committee for Guidance on Setting and Enforcing Speed Limits. A registered professional engineer in California, he holds a bachelor's degree in civil engineering from Sacramento State University and a master's degree in public administration from San Diego State University.

DAVID K. WILLIS is currently the President and CEO of the American Automobile Association Foundation for Traffic Safety. He has also served as director of the ATA Foundation, Inc., and as the director of policy research for the Motor Vehicle Manufacturers Association. Mr. Willis is a member of TRB's Committee on Safety and Mobility of Older Drivers and a member of the Board of Directors of the National Sleep Foundation.